Successful professional reviews

for civil engineers

by H. Macdonald Steels

Thomas Telford

Published by Thomas Telford Publishing, Thomas Telford Services Ltd, 1 Heron Quay, London E14 4JD

First published 1997

Distributors for Thomas Telford books are
USA: American Society of Civil Engineers, Publications Sales Department, 345 East 47th Street, New York, NY 10017-2398
Japan: Maruzen Co. Ltd, Book Department, 3–10 Nihonbashi 2-chrome, Chuo-ku, Tokyo 103
Australia: DA Books and Journals, 648 Whitehorse Road, Mitcham 3132, Victoria

A catalogue record for this book is available from the British Library

ISBN: 0 7277 2613 7

Typeset by MHL Typesetting, Coventry
Printed in Great Britain by The Cromwell Press, Melksham, Wilts.

'My task which I am trying to achieve is, by the power of the written word to make you hear, to make you feel — it is, before all, to make you *see*. That — and no more, and it is everything.'

Joseph Conrad, *The Nigger of the Narcissus*
from *Four Tales*, Oxford University Press, 1954

Acknowledgements

The advice in this book is based on the experiences of well over a thousand candidates, who have freely shared their doubts and fears, their jubilation and despair, their frustrations and delight, since I became the Institution's representative in the North-West in 1988. It also leans heavily on the many frank discussions I have had with the Reviewers over the years, both during the review processes and at many seminars where we have cooperated to explain the system to prospective candidates. In particular, I am indebted to the perceptive advice of Rodney Day at a critical stage in the book's development. I have also been privileged to be at the heart of the Institution during an exciting period of rapid change in the entire industry, a period which is not yet complete; I have thus been party to many of the discussions, debates and arguments behind changes in the procedures and criteria for the Reviews. Notwithstanding all this, I take full responsibility for the opinions, advice and guidance contained within these pages, all of which is given in good faith but without prejudice.

Preface

The outcome of *Effective training for civil engineers* should be *Successful professional reviews*, so there is a basic logic in writing this sequel to my earlier book. Nevertheless, it may surprise some people that I consider a further book on the Reviews to be necessary at all. After all, the Institution publishes full guidance in ICE 101 and 102 and there is already one book available – *Preparing for the professional reviews of the Institution of Civil Engineers* by J. Venables. However, like its predecessor, this book provides answers to the questions I am most commonly asked by candidates, questions which would not be asked if existing coverage were adequate.

This book explains the reasons behind the various parts of the process, because it is my experience that knowing these helps engineers to understand what they are trying to achieve, have clear targets at which to aim and hence to focus their whole effort on the efficient achievement of the objective – a successful review. The guidance is largely independent of the rules; readers are assumed to have copies of the current Institution of Civil Engineers documents and to be entirely familiar with the content. The book can also be read independently of *Effective training*, although the skills and abilities which are now to be demonstrated must have been developed through adequate training and experience, properly utilised.

You will quickly realise that I cover both the Chartered Professional Review and the Incorporated Professional Review, because they are complementary and both test the abilities of professional engineers. This is contrary to a widely held view perpetuated in the industry and, indeed, in academia, that Incorporated Engineers are somehow second-class citizens – technicians, useful but not 'proper' engineers. This view, which is contrary to the information in ICE 101 and 102, is thoroughly demolished in Chapter Two, but no doubt will still take an inordinate amount of time to be dispelled completely. At the time of writing, further strenuous attempts are being made by the Engineering Council to reaffirm this interpretation and at last to impose the standards first promulgated over 25 years ago; standards which relate closely to those agreed under the European General Professional Directive for Group 1 Engineers throughout the Union.

The debate which is proceeding apace at the moment is not about the necessity of improving standards, which seems to be a general concern, but only about the best way of achieving this. In anticipation of a satisfactory system being agreed, I have differentiated between the two kinds of professional engineer only where necessary. Most of the advice is common to both the Chartered and Incorporated Professional Reviews, but where there is a difference, separate divisions are clearly made within the chapters.

All the examples are taken from real submissions and events. As a matter of principle, only those perpetrated by *successful* candidates have been used, but even these have been 'doctored' to prevent any correlation of an example to a particular candidate. You may well think that you recognise some of them – one delegate at a seminar said that he was going home to rewrite his Reports, which he admitted contained nearly all of the examples I had quoted!

H. Macdonald Steels

Contents

Chapter One

Background to the role of the professional civil engineer

It is very easy in the hurly-burly of everyday life to forget the overriding philosophical nature of our work. The monarch, in the Institution's Royal Charter, states emphatically that 'many important and public and private works and services in the United Kingdom and overseas which contribute to the well-being of mankind are dependent on civil engineers'. Our profession is thus acknowledged as being at the very heart of civilised society. Far too many of the young entrants to our profession do not seem to lift their hearts and minds above the mundane day-to-day activities which occupy them to realise what a profound and far-reaching effect every aspect of their work is having.

Whatever we do, from the largest dam to the smallest patch repair, affects people and the environment. We all seem to remember that part of our Royal Charter which originally defined the profession as 'The Art of directing the great forces of power in Nature for the use and convenience of man', but how many of us have read on, to the statement that our job requires 'a high degree of knowledge and judgement in the use of scarce resources, care for the environment and in the interests of public health and safety'. This, to me, is the most succinct summary I know of the true nature of our business.

Every candidate must understand that the Institution is required by its Royal Charter to 'ascertain persons who by proper training and experience are qualified to carry out such works'. All these rather grand ideas may seem a million miles away from the day-to-day problems faced by civil engineers, but each one of us must never lose sight of these ideals and must seize every opportunity to express views and opinions based upon them.

On behalf of its members, the Institution is already taking a leading role in the debates on the environment and what has become known as sustainability, 'meeting the needs of the present without compromising the ability of future generations to meet their own needs' (The World Commission on Environment and Development, 1987). Note the importance of the term 'needs', not wants or desires — a matter of judgement and persuasion — no longer 'predict and provide' but 'target and manage'. Sustainability is a global concept both in absolute terms and in the context of sharing resources. All must act in concert, something which thousands of years of history suggest is not possible. But there are encouraging signs that the world's societies are realising that individual elements cannot act in isolation. More than ever before, compromises are being sought worldwide, to stop warfare and combat sickness and famine — perhaps not particularly effectively to date, but as yet there is not a lot of expertise.

Civil engineers, with their unique set of skills, particularly their ability to take the broad view and seek acceptable compromise, must play a major role in the entire process. Much of our society has vested interests — what President David Green described as 'single issue politics' — giving rise to the uncompromising pressure group. Civil engineers should rise above these vested interests, to utilise scarce resources, care for the environment and protect the safety and health of the public. The Institution is intent on protecting this high and responsible ethic and its Review system reflects the global responsibilities which this entails. You are therefore unlikely to have a successful review if this high ideal is not a motivating force behind your preparation, submission and performance on the day.

Engineering the environment

The concept of a need for the United Kingdom to improve or protect its environment did not actually arise until the early

1800s. The dramatic effects of the Industrial Revolution quickly made concerned people realise that there was a necessity for legislation to limit the physically dangerous and harmful effects of many manufacturing processes and to reduce the risks inherent in a rapidly increasing population migrating into the new towns (the population of the UK in 1800 was 10.5 million, compared with nearly 60 million today).

During the mid-nineteenth century an enormous surge occurred in the work required to build an infrastructure capable of realising people's expectations, as enlarged by the revolution and defined by the new laws. This led to the great sewerage and water supply schemes, improvements in transportation and the development of strong municipal authorities. This was a supreme opportunity offered to the embryonic British civil engineering profession and they responded wonderfully − with vision, creativity, ingenuity and courage.

The current flurry of fundamental change prompted by the release of market forces in this country, coupled with the technological revolution, is already beginning to spawn another burst of legislation to rectify the new social and environmental problems it is creating, just as after the earlier Industrial Revolution. This time around there will not be the same scope for massive infrastructure works; we are reaching saturation point in this relatively small island. But already there is perceptible pressure for improvement and rethinking of the very basis of our existing infrastructure, pressure which will grow until the pendulum swings in the other direction yet again. At that stage, which in my view is not far away, there will be an enormous need for highly competent and skilled engineers, not to continue to extend, but to make the best possible use of the existing infrastructure, with relatively minor capital works to improve its efficiency, and to devise innovative and ingenious solutions to the many problems.

Meeting the needs of today without compromising the needs of future generations requires a different outlook, altered skills and all the vision, lateral thinking and ingenuity we can muster. Our qualification system is already going through the painful and difficult transition to provide for this growing need. A successful

Review therefore requires candidates to demonstrate that they have the capacity to rise to this huge challenge.

Detailed changes to the review system will, and must, continue, to ensure that those becoming professionally qualified meet the developing needs of our industry and the demanding society we serve, but the fundamental philosophy behind the Reviews (as specified in our Royal Charter) will remain constant and must be kept in mind throughout.

How long will it take to prepare for the Review?

The Professional Review is the culmination of many years' work and personal development, which started long before you even went to university or college. It is the point at which many strands of development are brought together – technical knowledge, professional responsibility, personal characteristics – to demonstrate and prove your competence to fulfil the role as defined above, within your particular field of expertise. It is most certainly not an examination of whether you are capable of doing your job, which is unfortunately how many candidates (and some of their sponsors) seem to view it.

So one answer to the question is 'many years'. In reality, however, I hope you are continually reviewing your progress against defined targets – initially, your training objectives and, latterly, the available descriptions of what has to be demonstrated at the Review. There will come a point (usually quite suddenly) at which you feel confident about your ability – that is the moment to start preparing the submission. From this moment, and based on watching many candidates, I recommend a minimum period of some six months before the submission date. On this basis, a rough programme would be as follows.

Submission by	Draft reports by	Sponsors approached	Support documents by
15 February	end September	end November	end January
15 August	end March	end May	end July

Such a programme contains adequate leeway for the inevitable delays in responses from your sponsors and advisers and for the Christmas or summer holidays. But it still means that preparation has to be pursued diligently and constantly if the submission is to be of the highest standard, demonstrating your capabilities to best advantage. I have known successful candidates who started long after the dates suggested above, but they virtually abandoned everything else in a frantic surge of 'accelerated working'.

In addition to the preparation of the submission, you will also have to programme into your schedule the preparation for the interview, including a presentation, and the written work. It seems that few young engineers encompass these wider issues in the normal course of their work or by research and reading around their profession, so it is usually tackled like some kind of crash course at the end. I deprecate this, but acknowledge that this is the reality of most situations. However, do not underestimate the work involved. It is unlikely to be successfully squeezed into the two and a half months between the submission and the interview.

The jigsaw concept

Helping my young grandson to build a jigsaw, I realised that we were doing exactly what every candidate must do – construct a jigsaw. Each piece had to be examined in detail to see how the part picture on it interlocked with the pieces alongside. Continual reference had to be made to the picture on the box to see the position of each piece in the complete picture. This process mirrored precisely the system which I recommend for compilation of your Review.

All the components of the Review, the documents making up the submission, the interview and the Written Test or Essays in the afternoon, are each an essential part of the whole picture; they form the pieces of your jigsaw. All the pieces are important; none can be omitted. Everyone knows how frustrating it is when one final piece of a jigsaw is missing or mangled; you must not leave any possibility that the Reviewers might feel that frustration!

The picture is one of two for everybody, either AMICE or MICE, but the individual pieces are different shapes and sizes for each individual – no two jigsaws will ever have the same shape or size of pieces. This is why it is no good copying someone else's submission, it probably will not work for you. It is also why the

Institution is very reluctant to publish examples of successful work, either for the submission or in the written examinations; there is always a great danger that any published example will be perceived as the approved format – the 'formula for success'. This danger exists within organisations as well, where candidates are tempted to copy an earlier successful submission in the belief that this is the way to succeed – many have had a nasty shock!

So the picture for a Chartered Engineer is:

While that for an Incorporated Engineer is:

What you need to do is ensure that *your* picture is complete, with each piece locked into the others and with no gaps or overlaps anywhere.

The two jigsaws have the same number of pieces, with variations in only two of them. For a Chartered Engineer it comprises:

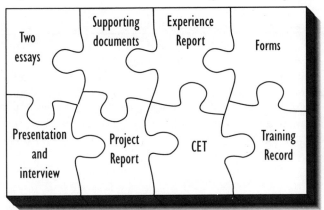

While for an Incorporated Engineer it comprises:

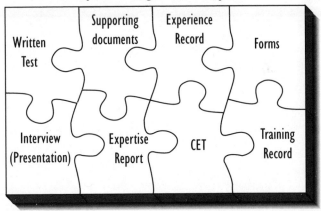

It always worries me when I am told, 'I've finished my Experience Report, now I'm going to start my Project Report'. Just as with a real jigsaw we compile different groups of pieces as we recognise them, so, in my view, it is good tactics to develop all the pieces of the jigsaw together, so that they are complete in themselves and form a complete whole, interlocking without any overlaps. Obviously, at the stage of preparing your submission, you will not know what is going to happen at the interview, but you should, even at this early stage, have in mind how you might respond to the invitation to 'take us through your submission', either formally for the presentation or, as often happens where it is not a stated requirement, informally as the opening of the interview.

Some components (the Training Record and requisite Experience) may have taken many years to compile, others may take only an hour or two, but represent many years of development. Subsequent chapters of this book consider each piece in turn and how it interlocks with others, but the next chapter spends some time looking at the completed jigsaw – the picture on the lid of the box containing all the pieces.

Chapter Two

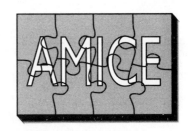

The complete picture

Before considering what the picture actually looks like, I think it is important to think about the underlying purpose of the Reviews. This is not some sort of one-stop examination, for which you can cram and swat and then forget what you have memorised a day later. It is a *review* of the benefits you have gained from the experience you have had and an assessment of whether these benefits have developed the required understanding and skills — it is a review of what you have become, not of what you have done. Obviously, one cannot truly be separated from the other, but the focus is very definitely on you, not your work. Your experience and specific projects are the vehicles by which you demonstrate your abilities; they will not, in themselves, impress the Reviewers. There is no automatic correlation between the prestige of the work and the benefits you gained.

What is the purpose of the Reviews?

I make no apology for returning to trying to explain the purpose of the Professional Reviews — both the Chartered Professional Review (CPR) and the Incorporated Professional Review (IPR) — an explanation which also formed part of my book *Effective training for civil engineers*. Despite the best efforts of the Institution and the Regional Liaison Officers, including myself, there still appears to be widespread misunderstanding, exemplified by comments I continue to hear.

There is nothing mystical about the Reviews; the present system is the culmination of over 50 years' experience in the review process. Further changes will be made as more operational experience is gained and as the competencies being reviewed become more clearly defined. I sincerely believe that the present review system is the best yet — by no means perfect, but better than any previous system. However, two problems will always be inherent:

(a) the personal qualities and understanding being sought are intangible and therefore difficult to define,
(b) young engineers dominated by (and highly proficient in) very structured examination systems will always seek syllabuses, model submissions and defined targets. None of these can be produced.

The Professional Reviews are designed to enable you to demonstrate that you have become a professional engineer, not just list the experience you have had. This begs the following fundamental question.

What is a professional engineer?

I included four 'definitions' of a Chartered Engineer and three of an Incorporated Engineer in my book on effective training; all are widely available in publications from the Engineering Council, the Fédération Européenne d'Associations Nationale des Ingénieurs and our Institution (the Chilver Report). All date back to the 1970s, so there is nothing new in them — I merely brought them together. In fact, engineers have been attempting to 'define' themselves since long before Henry Palmer and Thomas Tredgold in 1828 — even Aristotle had a go!

We all remember Tredgold's 'the Art of directing the great sources of Power in Nature for the use and convenience of Man etc.' but how many recall that other part of our Charter, '... calls for a high degree of professional knowledge and judgement in making the best use of scarce resources in care for the environment and in the interests of public health and safety'? Yet every candidate professes to have achieved Objective 1.1 or Activity 0.6. I now have a collection of 'definitions' from all over the world; all are trying to say the same thing, but none is a prescription. If there were an easy definition, then all the prolific writers on management, let alone engineering management, would have long since run out of ideas!

The only readily available information is contained in the booklets *Routes to Membership*, ICE 101 for Chartered and ICE 102 for Incorporated Engineers. While these are by and large 'method' specifications, they do include a brief coverage of the 'performance' expected, currently in paragraphs 5.4 and 4.4 respectively, where it states:

'The review procedures are designed to enable candidates to demonstrate that during their employment they have:

(i) developed and proved their technical and professional competence, including the exercising of engineering judgement requiring both responsible experience and the application of engineering principles; and

(ii) acquired an understanding of financial, commercial, statutory, safety, management, social and environmental considerations'.

Compare paragraph 5.4 of ICE 101 with paragraph 4.4 of ICE 102 − not a lot of difference! Surely this must convince you that an Incorporated Engineer is very definitely an engineer, not a technician? Few organisations, their management or, importantly, their staff structure yet recognise this.

These two key sentences encapsulate the whole purpose of the Reviews; every part of your effort should have these paragraphs in mind − your experience, training, Continuing Education and Training (CET), submission, interview and Written Test/Essays.

How do I decide that I have become a professional engineer?

It is evident that many engineers, both candidates and senior managers, have still not come to terms with the fact that the Incorporated Review now encompasses the more technical part of the former Chartered Engineer spectrum. Too many CPR candidates are still coming forward attempting to prove that they are in fact IEng, while the myths and misunderstandings arising from the old examination system of PE 1 and PE 2 live on, despite the Institution's best efforts. I still hear that 'you need a design for your civils', that 'you need twelve months on site/in design', that 'you need a bill of quantities and a rate build-up', that 'you can only count civils meetings if you have written 500-word reports', that you need to perambulate around all the departments in an organisation before the Training Agreement can be signed off. You may require any or all of the above for your development, your Supervising Civil Engineer (SCE) may insist, but they are not *requirements* – there are very few 'rules'.

What is required is that, somehow, you gain adequate experience to develop the characteristics and understanding being sought. The end product is 'defined' (albeit very loosely), but not the precise means of achieving it, apart from some very rudimentary Objectives or Activities. Of themselves, these will not be enough, but they do form a solid foundation upon which 'responsible experience' can be built.

But I am a specialist!

I can assure you from watching many engineers progress successfully through the Reviews, that it is possible to gain adequate experience in the most unexpected situations. The Institution no longer has routes for so-called specialists; at last it is acknowledged that every successful candidate is a specialist, whether in reinforced concrete, bitumen, geotextiles, groundwater movement, North Sea oil, traffic management, even as far removed as insurance assessment and actuarial work – but all are able to display the characteristics and understanding required. There can be no specification for their development, because everybody starts from a different point, has different innate abilities, differing education and different experience (even if working in the same office).

I have never had a Training Agreement

At least one trainee I know has never had the opportunity for a Training Agreement; his employment consisted of a series of temporary contracts (some of only a few months' duration) with organisations not on the ICE Index of Employers Approved for

Training. Yet he had an exemplary Training Record in ICE 107. How? Because the purpose of training and the required end result were fully understood and senior people had been cajoled into assisting. The record contained a complete set of critically annotated Quarterly Reports, a ratified set of achieved Objectives and more than adequate confirmed CET. Furthermore, all the required abilities were evident at Review.

This one example implies that all the information needed *is* available – but still the hearsay and half-truths remain. It is difficult to understand why, except that most people are far more comfortable with a set of rules which can be ticked off as completed, thus avoiding the need for judgement and assessment against a series of rather vague criteria. What each individual must do is continually assess themselves against the end-product, identify any deficiencies and then seek the experience or CET which will enable them to rectify their shortcomings.

What do the key paragraphs in ICE 101 and 102 actually mean?

To return to those key paragraphs – 5.4 and 4.4; at first glance they do not seem too onerous, but like all specifications they should be read with the intention of finding out how best to comply, i.e. achieve the performance, rather than trying to define the method. Many candidates try to emulate someone else's successful submission – it won't work!

Separating out the keywords reveals ten areas of competence. Defining each reveals the full breadth and depth of ability sought. Do note that the first subsection requires you to have developed and *proved* various things, while the second requires that you have developed an *understanding*. So you must have direct hands-on experience of the first group, but could have developed the necessary understanding of the second group by being curious, asking questions, discussing with senior colleagues and reading avidly.

You can decide your own definitions, but here are some suggestions which should help you to appreciate the full implications of these few words. Remember, they apply to both kinds of professional engineer.

DEVELOPED and PROVED

- *Technical competence:* the ability to apply established analytical methods and procedures in the context of a thorough understanding of the engineering principles upon which they rely, such that the procedures will not be used inappropriately.

- *Professional competence:* the exercise of professional skill and judgement with integrity and to the best of their ability to safeguard the public interest in matters of safety, public health and the environment and to uphold the dignity and reputation of the profession.

- *Engineering judgement:* the ability to go beyond established techniques, proven methods and documented precedents to develop a solution to a problem and to take personal responsibility for the effectiveness of the solution.

- *Responsible experience* (defined by the Engineering Council in *Standards and Routes to Registration — SARTOR*): the third Stage in the formation of an engineer beyond structured training (Stage 2) after academic education (Stage 1), in which 'the full required range of skills and abilities are developed and increasingly used under decreasing supervision' — 'to equip individuals as fully as possible for their first substantive post' so that they:
  - 'approach it with confidence'
  - 'do themselves justice in it'
  - 'make their best contribution to the success of the enterprise'.

- *Application of engineering principles:* an understanding of the fundamental properties of materials, basic behaviour of engineering systems and the imagination to anticipate possible loading conditions, thus enabling accurate judgements to be made about suitability, risk, safety and the proper use of resources.

Acquired an UNDERSTANDING of

- *Financial considerations:* the use, control and documentation of Money as the key resource, all other resources being dependent on it, namely Machines, Materials, Manpower and Methods (the 'five Ms').

- *Commercial considerations:* the balancing of costs and benefits in a professional manner to protect and benefit those to whom the organisation is responsible (e.g. clients, subcontractors, general public, staff, shareholders). The equity, working capital management, forecasting and presentation of accounts (profit and loss, balance sheet and cash flow) of the organisation.

- *Statutory considerations:* the entire legal framework within which the enterprise is operating.

- *Safety considerations:* the anticipation and management of hazards and risks in investigation, design, construction, use, maintenance and demolition within criteria defined in law or otherwise established as best practice.

- *Environmental considerations:* the management of resources in a project to maximise the benefit while minimising the disbenefits to the environment, this being the aggregate of all the external conditions and influences affecting all forms of life on this planet.

All these criteria enlarge on the Institution's Royal Charter, which all candidates profess to have read and understood to satisfy Core Objective 1.1 and Sector 0 Activity 1:

'a profession which calls for a high degree of professional knowledge and judgement in

making the best use of scarce resources
care for the environment and
the interests of public health and safety'

(paragraph 2).

The keyword implicit throughout all these descriptions is *judgement*.

So what is the difference between CEng and IEng?

There is a clear distinction between a Chartered Engineer and an Incorporated Engineer, but it is not the preconception which is common in the industry, which still equates the latter with 'Technician'. The Engineering Council's own descriptions show this absolutely clearly, but neither this body nor the many

Institutions awarding the qualification have been able to get this message accepted.

Incorporated Engineers are without doubt engineers, not technicians, who are experts in their particular field. Their role demands a practical approach, considerable technical competence and some managerial expertise and control in a particular aspect of the construction industry, with an understanding of the whole procurement process and an appreciation of the social, economic and environmental impact of their involvement.

Chartered Engineers, on the other hand, combine a thorough understanding of technical principles with broad, multidisciplinary professional and leadership capabilities to enable them effectively and safely to direct, change and progress the infrastructure and built environment by balancing financial, social, environmental and political implications and the effective and beneficial management of resources.

That is why the two ICE booklets, ICE 101 and 102, cover the same range of expertise in paragraphs 5.4 and 4.4 respectively. It is the level to which each of those items in the list is achieved which distinguishes between the two sorts of engineer. Generally, engineers do seem to divide into different kinds of people. Some are at their best 'doing' the technical engineering; they consider that persuading politicians and a sceptical public of the legitimacy of their proposals or seeking permissions and managing resources is frustrating, or they may not, so far, have had the opportunities to operate at these interfaces. Others revel in the persuasion and conflict of trying to persuade the public that what is being proposed is the best compromise solution, or of working in and organising a multidisciplinary environment, but retain sufficient technical understanding to realise the full implications of what they are doing.

An indication of the differences is given by the IPR requirement for ten Activities to be demonstrated to level B, that is a high standard of expertise in a limited range of work, where the person could, if necessary, supervise others doing similar tasks. The ten represent considerably less than 10% of the total number of Activities, i.e. the candidates are experts in a particular field and can demonstrate all the requirements of paragraph 4.4i. But candidates are also expected to understand the whole context of their work (for paragraph 4.4ii). So, dependent on the particular area of expertise, their profile might look something like this:

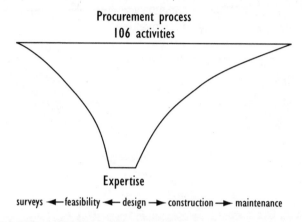

Procurement process
106 activities

Expertise

surveys ◄—feasibility ◄— design —► construction —► maintenance

Candidates for CPR, on the other hand, are expected to have a greater depth of understanding of the whole procurement process, gained partly by reading widely and partly by experience: for example, the mandatory Core Objectives embrace a wider spectrum of the process than might the ten Activities. So CPR candidates would not perhaps have such a depth of expertise in their particular specialisation; their profile would be quite different and might look something like this:

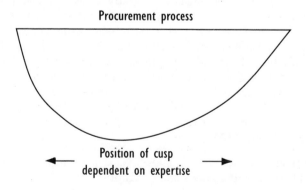

Procurement process

◄—   —►
Position of cusp
dependent on expertise

Obviously theses profiles are a gross over-simplification of reality. They do in a sense divide the world into racehorses and cart horses – both good at their job but capable of doing the other job, perhaps not particularly well (although I quite fancy the idea of a cart horse race). Where do eventers, steeplechasers or hurdlers fit in? Somewhere in the shaded envelopes?

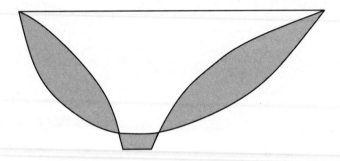

More difficult to place is the highly developed dressage horse? Your profile probably veers like a jagged saw from one side of the envelopes to the other, dependent on your particular experiences. So the reality is not quite as straightforward as my profiles suggest, but they do give an indication of the basic difference between the two sorts of engineer.

Generally speaking, most engineers make the transition from an essentially technical education to Chartered engineer via the specific expertise of what is currently called an Incorporated Engineer, and many now follow this path officially, by taking IPR first. I personally believe that a much better name would be that used in mainland Europe – Technical Engineer. As changes begin to take effect, I believe this progression will become the norm, and the majority will gain recognition as professional engineers at the time they become Incorporated Engineers. Once opportunities become available, some will broaden out to exert a significant influence beyond a strictly civil engineering context and will become Chartered Engineers.

The timing of these opportunities will depend to a considerable extent on the expectations and needs of the employer and on the developing capabilities of the individual. Some will have been recruited specifically to fast-track into this kind of role, others will gradually move into it, while some will prefer to become experts in a technological field. One route is not better than the other, just different. One problem at the moment seems to be that some pay scales do not accept these differences or recognise the value of having skilled individuals forming a suitably mixed team.

Am I a professional engineer?

Not until you are personally satisfied that you fully comply with one or other of the above descriptions should you proceed any further. I cannot to see how you can possibly convince others of your capabilities unless you truly believe that you are capable yourself. The preparation, drafting and collation of a submission is a long and laborious process, not to be undertaken lightly and surely not without a reasonable chance of success. The idea that you should make an application merely because you are time-served must not be allowed to cloud your judgement. Until you honestly believe that you *are* a professional engineer of the appropriate grade, don't apply!

Chapter Three

Starting to prepare the submission

Once you have decided that you are a professional engineer, either Chartered or Incorporated, the next question you need to ask yourself is:

'How am I going to prove it?'

From this point on, everything you do, everything you write, every form you complete, all the supporting documents you collect, are all aimed at one definite target – proving that you are a professional engineer of the appropriate grade. You must ensure that your jigsaw picture is complete, with each piece locked into the others and with no gaps or overlaps anywhere. The first and perhaps most obvious point is compliance with the rules.

Making sure you comply with the rules

You need to make sure that your copy of the specification or rules is the most recent version. Once you have made the decision to proceed, contact the Institution and check; ask the Reviews or Training Offices for an up-to-date set of forms and documents for the Review you intend to take. After all, you would not commence a civil engineering contract without the latest set of documents and standards, would you?

Then spend some time reading the documents thoroughly. It never ceases to surprise me just how many candidates reveal during a discussion that they are unfamiliar with detailed points in the Institution of Civil Engineers' publications. It pays to make notes or annotate the document to make sure you are entirely clear about which parts relate to you and which are irrelevant. Then I suggest that you read and digest the parts you have highlighted as relevant and make certain that you fully understand what is required. If in doubt, make use of the Regional Liaison Officer network, now available worldwide through electronic mail.

Only when you are absolutely clear on how to set about the submission (and of course that you are prepared for it) should you start. One important part of the submission which usually gets left until too late is the bureaucracy – the forms.

Filling in the forms

A key component of the submission is the various pieces of bureaucracy. These too are, perhaps unexpectedly, a crucial part of the jigsaw. It is amazing just how many candidates fill them in incorrectly, perhaps because they seem to be left to the last minute. I presume this is because they are not seen as a particularly important part of the submission. Yet they can help significantly with the overall impression.

These forms will create an impression. Like all the other pieces, this piece of the jigsaw must be used to support your claim to be a professional engineer. Untidy or difficult to read submissions filled in with an inappropriate pencil (even if the information itself is correct) all conspire to raise initial doubts about your professional capabilities. Do not leave this task to the last minute and hurriedly complete them, only to have your submission returned as non-compliant.

The space for the synopsis of your training and experience is quite limited and requires careful consideration of the best way to summarise what, by now, is a significant period covering a great deal of experience, probably obtained in many projects. Obviously, all the factual information must be there, covering everything in which you have been involved, but it may mean that some projects are clumped together – for example, 'five rural water treatment works about £½M each'. Do not reduce the font to microscopic 6 point to cram more words into the allocated space. All you will be demonstrating is that you cannot decide what is critically important – and risking straining your Reviewers' eyesight (and patience)!

One of the forms asks you to state 'My submission relates to . . .' followed by a series of options. This grid is a match for that completed by the Reviewers, indicating to the Institution the subject areas that they feel capable of examining and is the method by which the computer ensures that at least one of your Reviewers is familiar with your sphere of work. As a manual check on this match, you are also asked to submit a synopsis of your Project Report 'not exceeding one A4 page', by which staff who know the Reviewers can check that at least one of the chosen pair is sufficiently familiar with your work. So the synopsis may only be a short paragraph if that is all that is needed to give an indication of the type of work covered. Again, once you know the purpose, what needs to be done becomes more obvious. I have heard of people taking hours to write a comprehensive synopsis because they mistakenly compared it with that needed for a thesis.

Choosing your sponsors

Your sponsors should be chosen with care, not just a collection of available people from within your office. They do have to answer some fairly searching questions which you should read before asking people to commit themselves. It does you no favours for the Institution to be told, 'I am not in a position to answer this question' or 'Others may be better able to answer this question'. One sponsor should normally be the person who trained you, your SCE if trained under Agreement, but if that Agreement ended some time ago and you lost touch with them shortly after the Training Review, are they now in the best position to vouch for you as a professional engineer? Such a shortfall would need to be covered by your choice of the remaining sponsors.

While not a requirement, I think it appropriate to ask your sponsors to initial alongside those parts of your experience of which they have direct knowledge — first, it authenticates that experience and secondly, it demonstrates that they have seen and read that report and should have decided whether they think you have a reasonable chance of success, something your Lead Sponsor is required to have done.

Do not necessarily look for sponsors solely within your own organisation. Why not ask someone from your consultant/contractor/client/promoter to signify their respect for you as a professional engineer? While you may have had contractual differences and disagreements, surely the Reviewers will be favourably impressed by their opinion of you as a person? Think about the impression you are making all the time — the *purpose*, and choose your sponsors accordingly.

The Lead Sponsor

You have to nominate one of your sponsors as the Lead Sponsor; it should intrigue you as to why. The reason is that the Institution has become irritated by people sponsoring candidates who are totally unsuitable, often without even seeing the submission, and wishes to be able to reprimand those who do. Sometimes this results from an inaccurate or outdated perception of what the Institution requires of candidates; by putting the onus very much on the Lead Sponsors, the Institution expects them to familiarise themselves with the current requirements.

So your Lead Sponsor has serious professional responsibilities; to be properly accountable to the Institution for their opinion of you, they must therefore be a member of an appropriate grade. Their responsibilities are threefold:

(a) to satisfy themselves that you have a reasonable chance of proving you are a professional engineer,

(b) to certify that your submission is a reasonable reflection of your experience and capabilities,

(c) to check that none of your other sponsors has misgivings either.

If this relatively new alteration works properly, then the pass rate should rise.

When should the Reports be ready?

One implication behind the responsibilities outlined above is that before your sponsors (and particularly your Lead Sponsor) can complete their forms, you must have your Experience Report and your Project or Expertise Report practically finalised. Bearing in mind the logistics of obtaining the cooperation of your sponsors, sending out the forms, completion and return to the Lead Sponsor, who puts them in a sealed envelope to send with your submission, the date for completion of this aspect should be about one month before submission. So do not delay! I despair of the number of candidates who apply to sit their Review in the provincial centre last in the list 'to give themselves another month for preparation'; it is hardly surprising that their names do tend to appear in the 'unsuccessful' results list.

Chapter Four

The Training Record

Why should I submit a Training Record?

This question is regularly asked, particularly when the person has passed a Training Review or Experience Appraisal. There seems to be an impression that 'this bit is completed' and that therefore it should not be tested again. This is a complete misunderstanding! The professional reviews are holistic, that is, they cover *every* aspect of your abilities, knowledge and skills – there is absolutely no allowance for any previous review of progress towards this final goal. It is also important to remember that your Training Review or Experience Appraisal did not necessarily review the full depth of technical understanding required by a professional engineer; it merely ensured that at that time in your development, you were on course. That review was a staging post, not the destination.

In any case, there are bound to be parts of the jigsaw which do not appear in any of the pieces you have constructed; where, for example, will most people be able to show that they know something about the organisation of the Institution, or that they have an active interest in their chosen profession? It is unlikely to feature in either Report or in the supporting documents. And there are other examples, particularly centred on Core Objectives 1 to 9 and Sector 0.

What if I do not have a Training Record?

Even for those coming in on the Direct Entry route or the Senior Route, or via an Experience Appraisal, any of whom may not have a formal record of training, I would expect a short record. It might only contain a few meeting reports; there are unlikely to be any Quarterly Reports, but there could well be reports on particular parts of your experience which were significant in developing your skills and abilities. In any case, the record is an ideal opportunity to demonstrate your understanding of how the Institution operates and how your own organisation is organised and financed, none of which is likely to be covered anywhere else. It may also be appropriate to cover any topical concerns, particularly about education and training or technical and environmental matters of direct concern to your part of the procurement process.

The Training Record is another opportunity for you to put before the Reviewers the development of your expertise; it seems silly to forgo it. This is most important if you are using a subject for your Project Report which does not enable you to demonstrate all the characteristics and knowledge being sought, or where the ten chosen Activities do not show the entire range of your experience. If, for example, CPR candidates feel that their technical competence is underplayed in an essentially management-based project, then the opportunity is here to include purely technical support documents which perhaps were originally used to demonstrate achievement of Core Objectives 8 to 16. This particular documentary evidence could appear bound to the Experience Report, as allowed by paragraph C3.

An IPR candidate does not have the same opportunities in the current framework, where no supporting documents are allowed behind the Experience Report. Neither do the rules say that 'a Training Record is desirable but not mandatory', as they do for CPR. However, the rules do not specifically *exclude* a Training Record, so if you wish to put more of your experience before the Reviewers, submit one. But do make certain that it is compact and makes a significant contribution to the totality of your case. The Reviewers will not be impressed or swayed by quantity as a mask for inadequate quality!

There is one very important point to bear in mind, however. You must not rely on the Reviewers reading the Training Record. Your Reports, particularly the Experience Report, must stand alone, covering all your experience and expertise, and should neither refer to, nor rely on, the Training Record. Some Reviewers may only glance through the Record, a few may avidly read all of it, but nearly all will _look_ at it.

What are the Reviewers looking for?

The Reviewers will be looking for continuous and steady improvement, whether any reports were written in batches (which is easy to spot) or arrived more or less at regular intervals. They will not expect the first few to be particularly good; after all, you did not have much experience of reporting at that stage, but a significant improvement in deciding what to report and how to report it concisely and succinctly will be sought. They will see how quickly you adapted to a work environment and began to appreciate the wider issues beyond your immediate involvement. They will also get some idea of the support and encouragement you received from your Supervising Civil Engineer, Delegated Engineer or whoever it was who read your reports. To do all this, they will perhaps read little snippets, but in the main, many only read the written comments you received and assess your reactions to them.

You may have made a bit of a mess of your formal training. Do not be afraid to say so! Everyone makes mistakes — it is what you did about them that matters. No one is more aware of this than these very experienced Reviewers, all of whom will have done things which, with hindsight, they probably regret. There is a sheet included in most Schemes asking for your views on your training and experience — why was it put in there? Because the Institution wants you to commit yourself and tell them not only the benefits, but the downside. If things went awry, what did you do to rectify them?

I had one trainee whose Agreement lasted over six years. He admitted in conversation that it had taken some time for him to realise just what becoming a professional engineer actually entailed. When he eventually did, he took drastic steps to bring himself back on track. I advised him to write all this down in his comments and include a copy of the letter he had written to his SCE seeking his agreement to a proposed course of rectification, which included a timetable and projected date for a Training Review. He was very reluctant at first, but eventually did, and was very relieved when one of his Reviewers congratulated him at the start of the interview on 'the professional attitude he eventually took to his training'!

None of this assessment is going to fail or pass you, but it does give your Reviewers a little bit more of the total picture of you, your personal attitudes and your professional development.

Should I submit the complete Training Record?

When preparing the Training Record for submission, remove *all* extraneous material, anything which does not contribute towards your Review. This part of the package (sent to one Reviewer only) is in addition to the 1 kg limit. This does not give you carte blanche to put in as much as possible, but to take a responsible attitude and include 'just enough'. Remember, it is not a Training Scheme you are submitting, but a Training Record – what you actually experienced, not what you set out to do – the two may be very different! Copy reports back-to-back and retain the originals; this is a legitimate weight-reducing ploy, but photocopying on to airmail paper is again seen as a demonstration of your inability to decide what is important.

Remove all guidance notes and information for your benefit from the Training Scheme, leaving only that information which supports the objective of the submission. Remove any Specific Objectives or Activities which were not achieved and put in any which were not there originally. Of course, you cannot change or remove any of the Core Objectives or Sector 0.

If your Agreement was prolonged, so that you have over two or three years' worth of Quarterly Reports, you may feel it inappropriate to include them all. The temptation will be to leave out the ones which, in your opinion, are not particularly good; inevitably, these will be the earlier ones, which will probably now appear naive and rather inconsequential. Do not succumb! What your Reviewers would like to see is the whole spectrum of your development during that period. It is better to remove those reports which cover consolidating experience rather than new experience, which did not make a particularly significant contribution to your development as an engineer. Whatever you decide to do, attach a note to the front of the collection to explain how you made the choice, so that the Reviewers do not suspect (as otherwise they might) that you have left out the bad reports!

If you kept a number of photographs and other material such as calculations and brochures with your record, go through them in detail to see whether they will help the Reviewers — remember the overriding objective! If they are more an aid to your recall then remove them, but perhaps consider taking them with you on the day in case questions are asked which would be better answered by reference to further information. You should in any case, take the originals of all reports with you to the interview just in case there are any queries.

Bind the whole record together with your comments on your training experience at the front, followed by a one page summary of training; this enables you to draw the Reviewers' attention to any matters which you think they ought to be aware of — if these sheets are left somewhere towards the back, they may be overlooked.

Chapter Five

Continuing Professional Development

What is Continuing Professional Development (CPD)?

'The systematic maintenance, improvement and broadening of knowledge and skills, and the development of personal qualities necessary for the execution of professional and technical duties throughout working life.'

CPD is obligatory; Rule 14 of the Institution's Rules for Professional Conduct currently states:

'A member shall afford such assistance himself or through his organisation as an employer, as may be necessary to further the formation and professional development of himself and of other members and prospective members of the Profession in accordance with recommendations made by Council from time to time'.

Currently, the Council's recommendation is that every Member records an average of five days of CPD each year – obligatory but not mandatory. This requirement is in accord with nearly all the professional Institutions and is likely to increase in time. The Council of the Institution does not yet intend to make CPD mandatory, as this would involve expensive routine monitoring and legal sanctions, but it may become necessary. It is, however, *obligatory*, i.e. it is a professional, rather than a legal requirement.

Why is it called **CET** for trainees?

CPD becomes Continuing Education and Training (CET) for Student and Graduate Members, who are not yet professionally qualified and could not be disciplined if found to be cheating. Their CET must therefore be ratified by a Member, who is professionally accountable to the Institution, on pages 2 and 3 of the Record. After qualification, when you are accountable to the Institution yourself, CPD is self-certified on the remaining pages (no signature column).

Why bother?

The requirement was superficially viewed by many with alarm, particularly at the cost implications of 'sending everyone on courses' — not merely the cost of the course, but the productive time lost and expenses involved. There was initial resistance to what was seen as an imposition. I think most have now realised that this was a mistaken overreaction. It is, in essence, a system of Quality Assurance, a written audit trail of your continuing development.

There are further reasons for bothering:

(a) Although the Institution has not made it a mandatory requirement, evidence of CPD is required when you join (except currently for Engineering Technicians) and when you change level of Membership, notably from Member to Fellow.

(b) The Institution intends to carry out an annual check on a random percentage of its membership.

(c) Perhaps more significantly, consider the possible implications if you:

    (i)   give evidence in a legal/contractual case,

    (ii)   need Professional Indemnity insurance,

    (iii) are accused of professional negligence (a possibility more likely than hitherto given today's litigious attitudes),

    (iv) are required by a client to disclose the qualifications of staff you intend to use on a project as part of a tender proposal.

Would it not be reasonable for the lawyers, insurers or client to ask what has been done to maintain or enhance professional competence since qualification? How would your firm react if they were asked point blank, 'What right has this employee to be making such decisions?' and they were unable to *prove* (to the satisfaction of the risk assessor) your up-to-date knowledge and competence. Record what you are doing and you will not be caught out!

What does CET actually entail?

Since a CPD record became a requirement, engineers have found that they are in fact already doing at least five days per year and, in most cases, considerably more. Today's rapid changes, not only in technology and the legal framework, but also constant reorganisation and redeployment, mean that anyone who does not keep their knowledge up to date or learn new skills will soon be left far behind. Flexibility and adaptability are vital commodities in today's uncertain marketplace; for these, CPD is a prerequisite.

The list inside the front cover of ICE 108 covers all the methods of keeping up to date; these methods go far beyond 'going on courses'. It is possible to fulfil the minimum requirement without ever attending a course, although there are some things which are best covered in that structured environment. The only proviso is that, since the system is self-certifying, it is preferable for there to be some tangible end result, to use as proof if ever needed.

In practice, therefore, the CPD record is not an additional imposition, but a requirement that you properly record what is probably already being done. Many now produce Quality Assured projects, few staff are themselves properly Quality Assured. But isn't this the next logical step?

How do I decide what CET is needed?

CET is a method of continual self-development; it should be an intrinsic part of your continuing training and development, not something which is a 'bolt on' extra. It is necessary to routinely examine your current levels of skills and knowledge and measure these against those that already are, or probably will be, required in the future. This is not difficult if done in a rational and formal way. Many organisations now have formal systems for annual assessment of their staff and their training needs, but if yours does not, do it yourself.

The Institution has produced a self-assessment document to assist in the identification of essential CPD — *'Management Development in the Construction Industry'* available at modest cost from the Thomas Telford Bookshop in the Institution. I have, from personal experience, found it very easy to use and it certainly identified my needs! Candidates are increasingly including the detachable assessment sheets from this booklet as part of their written submission. It seems like a good idea to *demonstrate* clearly and very concisely how you identified the needs and what you did about them.

The key criteria are Continuing, Relevant and Progressive; going on a basic site safety course three years after first working on site is a sinecure — merely 'complying with the rules', neither progressive nor relevant and therefore probably unacceptable. While it will not be a reason for failure, it could well be held in evidence against you.

A strong indicator of Relevance for one critical aspect is supplied by the Core Objectives. The minimum number of days on safety training recorded for the Reviews was increased to three (for CPR) and two (for IPR), to cover the three aspects of safety — personal (Objective 7), in design (Objective 18) and in implementation or construction (Objective 27). For CPR, all three must be covered, for IPR (which is more specialised) the first and probably one other, depending on your area of expertise. However, these are bare minima; I personally would expect most candidates to have had more formal training in the fundamentally important matters of hazard and risk identification and management than would be represented by the minima.

The requirement for continuing and progressive education and training was probably the origin of the 'rule' that ten days should be spent on technical training (the early career), ten days on managerial (the latter stages of formation), and the remainder should be a mixture of both. Since most young engineers start by utilising the technical knowledge gained from their academic course and gradually move towards a more managerial role, this is probably what will happen, but it is not a requirement. For example, I know of successful candidates who are able to demonstrate that they had not needed any formal management training at all.

Also look carefully at the personal qualities being developed, which are necessary to many other professions; things like time management, team working, personal relationships, negotiating skills, running a small business, a foreign language. Consider material and courses beyond civil engineering – the local Chamber of Commerce, Further and Higher Educational courses (both vocational and recreational), distance learning material from other Institutions.

What the Reviewers will be looking for is a concerted and considered programme to put right areas of your expertise which you identified as needing attention. If you find using the English language difficult (and many engineers do) then have you tried to improve by attendance on a suitable course? If you have had difficulty with contractual disagreements or public meetings, have you been on a Negotiating Skills or Dealing with the Public course? If you went to work in a laboratory, did you first find out what the hazards were likely to be and how to avoid them?

Do keep in mind the Council's current recommendation for all of us – an average of five days per year. The minimum requirement for CPR is 30 days; for IPR it is much less, reflecting a concern that many employers are already showing considerable commitment in sponsoring most students on day-release academic courses. If your professional development has taken longer than the minimum, will achievement of the minimum requirement for CET be satisfactory? Again, of itself this would not be a reason for failure, but it is going to raise doubts.

CET is another and important part of the total jigsaw, not an appendage! It interlocks particularly with your Training Record as part of your overall development and will be consistent with your Experience Report.

Chapter Six

The Experience Report

The Experience Report is the fourth piece of the jigsaw to be considered and the first of the major reports. Like all the other pieces, it must not be written in isolation: you will be dependent on any records you have (see Chapter Four), one form requires a c.v. and you need to consider what supporting documents (Chapter Nine) you might need. It will also tie in with your Continuing Education and Training (Chapter Five). In addition, the Project or Expertise Report (Chapter Seven) will arise from one or more areas of your experience.

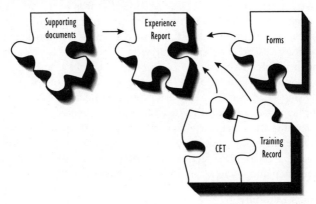

By now, you will have had considerable experience, probably something in excess of five or six years, covering a significant number of projects. It will not be easy to record all this experience in a mere 2000 words (CPR), or 1000 words (IPR) and, from personal experience, the Institution will not allow any dispensation, however many years' experience you are trying to describe. Yet everything must be included; you cannot discount experience, every piece of which makes us what we have become.

Most of us have difficulty in writing succinct yet comprehensive reports, partly because there is a growing tendency to believe that communication requires you to tell people everything immediately. How on earth did the historic Empire builders manage without fax, email and the ubiquitous mobile phone? The answer is that they knew how and how much to communicate and how to delegate and develop trust. Our society is bombarded with verbosity on all sides, particularly from television and radio, so we all tend to fall into the trap of using a paragraph where a sentence would do. This example (written by a Chartered Engineer) arrived on my desk shortly before I left my previous role; I have no doubt that everybody could cite similar cases:

> 'Reference my letter of late November, on the above subject, due to pressures brought to bear the situation has been eased and we can now go ahead with repayment works in 19??/?? for which orders have not yet been received.
>
> I would however remind you, if indeed that is necessary in the present climate, that because of stringent financial limits we are operating under our direct labour resources are consequently severely restricted. Repayment works in general involve us in much greater expenditure of our direct labour due to cabling and jointing operations, than the provision of ducts and jointing chambers etc, the latter being done by contract under our supervision. For this reason and because of difficulty in acquiring the necessary and in many cases special stores at short notice may I ask for your continued cooperation in taking due note of the time factors we state when furnishing you with estimates and quotes for projected schemes.
>
> Yours faithfully'.

The same information and the request for cooperation could be transmitted in two sentences:

'I am pleased that the problems outlined in my letter of late November have been resolved; we are now able to undertake orders for repayment works in 19??.

Your cooperation is sought in taking due note of the time factors in our estimates; this will minimise costs and avoid possible delays in materials supply'.

So keep a constant watch on irrelevant verbosity. Scrutinise every sentence. Is it necessary? Does it make a positive contribution to the overall objective? Does it use the least possible number of words?

The production of this Report requires you to be absolutely clear about exactly what you are trying to prove so that you do not waste any of the words. It will take time, help from colleagues and friends and a great deal of editing. Quizzing many candidates after their Reviews on what advice they would give to the next group, the answer is almost invariably, 'It takes twice as long as you expect!' In my opinion approximately four months is a reasonable and realistic period, although I well remember one candidate who successfully wrote both his Reports in a fortnight!

What is the purpose of the Experience Report?

The Experience Report piece of the jigsaw seems to me to be answering two fundamental questions:

(a) have you had adequate opportunities for sufficient experience?
(b) have you benefited adequately from that experience?

Consider the answers to each of these questions in detail.

Have I had adequate experience?

This first question is relatively easy to answer, since it is factual, based on a list of your appointments and the projects you worked on. But beware! A c.v. alone is not enough. What you must spell out are your precise responsibilities; for example: 'I was the sole Engineer's representative (ARE) on site' is a statement of fact but does not indicate the precise level of your personal responsibility. What responsibilities were formally delegated to you and how was everybody, including you, informed? Did the Engineer (or their Representative) come out every day to see how you were getting on? Did they come out once a week, or did you only approach the Engineer whenever you felt it necessary. All three situations represent completely different, and increasing, levels of responsibility, but any one of them is implied by the initial statement and so there is immediate doubt and uncertainty about your exact role.

'Setting out' is another common and bland statement which masks a multitude of responsibilities. You were probably personally responsible not only for making sure that the work being undertaken was in the correct position and of the correct size, but may also have had considerable responsibility for quality control of, for example, reinforcement, materials, properly tightened connections or the condition of ground samples.

The limit on words does not allow you much scope to answer both of my questions, particularly if your experience covers several years and many different projects. Since you are trying to prove that you have become a professional engineer and have covered an adequate range of experience, I recommend a greater emphasis on the latter part of your career to date, which is where you should have been deriving maximum benefit.

To help further in overcoming this problem of limited space, I suggest you write an introduction to include much of the factual information which will answer the first of my questions. The format which seems to work best is a one page, three-column foreword (not included in the word allowance) which looks like this:

Dates	Projects worked on,	Your actual
Name of company	with an indication of	responsibilities
Job title	size, complexity, cost	(in some detail)
Person to whom responsible	etc.	
(grade in ICE or other Institution)		

The tabular format enables you to reduce factual descriptions to gain space for the benefits. Consider the following:

'On commencing employment with Midshire County Council Highways Department, Design Services (Roads) Division, I was assigned to the Major Improvements Section and gained much experience working on several highway improvement schemes including:

(a) designing accommodation works for the Bugwash Bypass and compiling the accommodation works bill of quantities using the MICRORATE software,

(b) preparing vertical alignments for various bypass schemes,

(c) producing working drawings, schedules and contract documents for a section of the Feetwet Ferry Southern Primary Route Improvement Scheme − the working drawings were drafted on the Computer Aided Design system,

(d) designing preliminary horizontal and vertical alignments and junction layouts, using Department of Transport design standards and producing approximate cost estimates for a proposed bypass to Southwich Town Centre,

(e) preparing cost comparisons for two alternative junction improvement options.

For the majority of these schemes, once I had been briefed by the Project Engineer, I was responsible for obtaining relevant data from various sources, performing design and compiling drawings and documents'.

Most of this information could, with strict editing, be incorporated into a foreword table, leaving most of the space to enlarge upon the two unsubstantiated statements 'gained much experience' and 'I was responsible for obtaining relevant data from various sources' which, as they stand, are practically useless!

The foreword gives the Reviewers a clear skeleton of your career through which they can more readily assimilate the contents of your Report. But do try to keep to one page. Anything more is becoming an addition to the Report and should be an appendix, which is nowhere near as effective.

How do I *demonstrate* the benefit I derived?

A foreword table enables you to omit much of the explanatory text in the body of the Report, enabling you to concentrate on the benefit you gained from the experience. You can therefore omit such things as:

> 'I then transferred to the M6 motorway in Westmorland, where as a Site Engineer for a group of five structures (three bridges and two large culverts – total value £1.5M) around Grayrigg, I was responsible for . . .'

because all this information is now in the foreword. The sentence can now start, 'On the M6 I was responsible for . . .', leaving you with some 28 additional words to describe the benefit you gained.

May I use supporting documents?

One other way of saving words is to use diagrams and sketches; this avoids lengthy explanations of site layouts, locations or details. Here again, many candidates make the mistake of using contract drawings. I believe it is much better to use simplified extracts from them or the sort of simple diagrams favoured by those who prepare publicity material. After all, contract drawings are required to transmit detailed and accurate contractual information, generally far too detailed for the purpose of your Report.

Another example:

> 'I was temporarily seconded to this site as an ARE. The works at this site were part of several 'Advance Works' contracts for major improvement works to a busy road junction over and adjacent to the London Underground. These particular works consisted of the construction of a pedestrian/cycleway ramp and associated retaining wall from an existing bridge over the underground lines to an anchored sheet pile wall at the entrance to a future subway under a road adjacent to the works'.

Can you visualise the site? I happened to know where it was, and even then had difficulty! It is crying out for a diagram. Not a copy of the A-Z which again would have contained far too much detailed information, but an outline sketch. And look at the wasted words: surely a secondment can only ever be temporary. Also, there are no less than five references to 'the works' in only two sentences – the candidate obviously never read the Report aloud. A sketch and drastic editing could release many words to inform the Reviewers of how the candidate benefited from the experience at this complicated site.

Since the Reviewers have to assimilate the information quickly as an adjunct to your experience, they do not want to have to unravel a complicated drawing to glean the basic information they need. To assist them, I suggest that you put any pictures within the text, alongside the part of the Report to which they refer; modern word processing equipment makes this relatively straightforward. There is nothing more annoying than flicking back and forth between the text and an appendix, made even more frustrating if the appendix is an A0 drawing! This annoyance is compounded if the appendices are not bound into the Report, but are loose in the submission pile.

If there is no alternative to an appendix, then one great idea I have seen is to print the appendix on the A4 right-hand side of A3 paper, so that the picture can be folded out and viewed alongside the text for as long as necessary before being folded back in. But do make sure that the flap is not bound into the finished document!

There is another advantage in such an approach: colour photocopying is expensive, so pictures within the text should (unless expense is no object) be in black and white, but appendices, being isolated sheets, can be printed economically in colour. Do check the printing quality available – I have seen far too many colour reproductions where the quality is so poor that it is difficult to see any detail at all.

Chapter Seven

CPR Project Report and IPR Expertise Report

There are significant differences between the Project Report for CPR and the Expertise Report for IPR, so the two Reports are discussed separately.

CPR Project Report

The '4000-word Report'

If the Reviewers now know from your Experience Report that you have had adequate experience and that you have benefited fully from it, why then do they also require a Project Report?

In my opinion, this is to demonstrate that as a result of your experience, you have become, to all intents and purposes, a Chartered Engineer. In other words, this report must *demonstrate* either that you have or that you could (given the chance) readily and confidently take the responsibilities and display all the attributes of a Chartered Engineer.

In that case, it is not really a project report at all. It is a report on your involvement in a project. It 'adequately describes' your personal experience, your understanding and part in the decisions which were taken and your role and responsibilities as an engineering manager. It represents the culmination of all the experience (Chapter Six), training (Chapter Four) and Continuing

Education and Training (Chapter Five) which you have had to date:

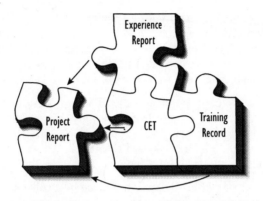

The Report should be 'well constructed' to demonstrate your formal communication skills. Pay close attention to grammar and syntax; again, get someone to read it who perhaps has a greater command of the language and listen to their queries and comments. If you have access to someone who does the organisation's marketing, ask them for their opinions; they usually come up with interesting ideas and thoughts. You do not need to act on all of their suggestions, but at least consider their very different points of view. This Report, like the Experience Report, should be clear and readable so that the Reviewers actually enjoy the experience and look forward to meeting the author.

How do I choose a suitable project?

This is, without doubt, the most common question I am asked about this aspect of the submission. There is still a residual belief among many engineers that candidates need a design, preferably in reinforced concrete. This is just not true!

The rules state that the project must be an 'adequate challenge' — technical, professional and managerial. How will you know that the one you wish to use is adequate? I suggest that rather than working from all of your experience and attempting to draw from it something which might work, it is preferable to work back from what you are trying to demonstrate and choose something which will achieve that aim. So start from the attributes of a Chartered Engineer; how can they each be demonstrated through the intended project? Use your Training Record and other accumulated documentation to provide inspiration and information. Make notes! It is surprising how easily ideas can be lost if they are not recorded at the time you think of them. This stage will take time, so do not leave it until the last minute.

If there is a choice of possible projects, this process will quickly reveal which is the best vehicle. Surprisingly, it is not always the most interesting or prestigious project which suits the purpose best. Very often, it is the awkward smaller project with a minimal management team which best demonstrates your abilities.

The Report need not be on one job — it can be a collection, but be careful in your choice. Remember, you are going to demonstrate *your* abilities and the more words used to describe projects, the fewer remain for this prime purpose. Generally speaking, more than two different projects become unmanageable, particularly in the short presentation, although again, this is not a 'rule' — one person successfully delivered seven projects in the fifteen minutes! If you are faced with this problem, then consider a generic project title, such as 'The assessment of highway bridges with particular reference to [and list the names of those to which you refer]' or 'Redevelopment of disused gasworks sites' or 'Improving the traffic flow in [a city]'. The projects can even be completely unrelated: 'Refurbishment of warehouse' coupled with a 'Cofferdam design', but link them by explaining to the Reviewers why both are included.

One thing which is still apparent is the overwhelming desire of some candidates to include a 'design' and a 'site' project. This is not a requirement, is not necessary and, most importantly, can leave you open to stiff questioning on the area of expertise which is less familiar to you. In other words, if your prime expertise is the management of site works, do not submit a design as the project! If the subject of the Project Report does not enable you to demonstrate your technical ability to best advantage, then use another piece of the jigsaw to cover the gap. The ICE documentation states that the Reports should contain drawings, numerical analyses and illustrations as appropriate — either or both can be supported.

The project need not be 'complete' i.e. from problem to hand-over, but it must have a clearly defined 'start' and 'end', for example 'Preferred route for Exchester Bypass'. In the lengthy progression of a highway scheme from identification of need to construction, the preferred route is a definable stage, so it does have a definite start and end.

There is also no need for the project to be particularly large or prestigious; the overriding requirement is that it must enable demonstration of your attributes as a potential Chartered Engineer – technical or managerial complexity are more important than prestige.

During the course of my years working for the Institution, I have seen many unusual projects used successfully by candidates. They include:

- Maintenance of the highway network in a National Park

- Management of the highway network in an inner city

- Installation and commissioning of a quality assurance system

- Improving the traffic flows in an urban area

- Installation of pay-and-display car parking in . . .

- Experimental testing of prestressed brick walls

- Tender buying.

All of these were successful – for that particular candidate. It is no good spotting one which might apply to you and then trying to make it work. These examples are only intended to give readers a view of the wide range and type of projects which could be successful – each candidate *must* choose something which suits their own experience.

All engineers tend to take what they do for granted. We tend to think that what we do is obvious or commonsense and forget the ramifications we went through to find a solution. It is therefore difficult to recall everything of relevance. Hence the importance of good diaries and Training Record and the need for time to unravel the original problem from your records and memory.

Once you have completed making notes on the individual Attributes, you will have at least sufficient material for 4000 words. The next stage is to organise the information into a coherent whole. Tackling it this way will ensure that most of the Report is devoted to your role, responsibilities, understanding and experience with minimum coverage of the project itself. In other words, you are targeting your report at what the Reviewers are seeking.

How can I minimise the word count?

To minimise further the 'waste' of words on explaining the project, make judicious use of sketches, maps and plans or photographs, each one chosen with care to illustrate exactly what you need. Too many candidates sprinkle photographs liberally throughout their Reports in a vain hope that they might impress the Reviewers. Beware! Of themselves, pictures will certainly not impress.

The ideal is to describe the project and the background to it in the first half (no more than two-thirds) of the first page, so that you can begin to describe your role within the first 100 words or so. This is not always possible, but it is a target. I saw the ideal recently, where a candidate was using a demolition project. The front cover showed a photograph of a tower block from the roof of an adjacent block; the proximity of other blocks and the infrastructure around was all too apparent, even the number of floors could be counted. The Report started 'My job was to demolish this block without undue disruption'. The back cover showed a pile of rubble, with the perimeter fence still intact! Very few candidates have this sort of luck, but it is something at which to aim.

Look very carefully at every illustration. First, to be sure that it actually does contribute towards the overall objectives, either demonstrating that you are a professional engineer, or reducing the word count. Secondly, look very carefully at the background to every picture: is there anything there which you would rather not show? Even experienced publishers have been caught out by pictures where something inappropriate is going on in the background. Too often, photographs open up discussions on peripheral matters like safety or efficiency which can take you unawares.

IPR Expertise Report

The '2000-word Report'

Choosing appropriate activities

For this Review, you are in the fortunate position of being allowed to choose the syllabus; you are required to choose ten Activities in which you must demonstrate total expertise, that is, at Level B of the BEKA system, so competent that you can supervise the Activity being done by others.

There are over 100 Activities. The first task is to sieve these down to those most relevant to you. Start by assessing your entire experience against the ICE Sectors and Activities. The Sector 0 Activities are compulsory and *additional* to the Activities chosen. Do not restrict yourself to those Sectors which appear to be most relevant; the list is beginning to show its age and its traditional arrangement reflects a system which is being changed by market forces, differing working practices and the drive for less confrontational contracts. This initial sieve will probably highlight at least 40 to 50 Activities from 4 or 5 Sectors and will give you a good indication of whether you have the required breadth of understanding (see Chapter Two).

The next stage is to extract from these Activities those which best describe your current involvement (although this is not a 'rule'), i.e. to produce a 'best-fit' current job description using the Activities. If you have a recently completed Training Agreement, these first two stages are virtually done when your Completion Certificate is signed and registered.

Why do I suggest you assess 'current' activities? For three reasons:

(*a*) you are probably a better engineer today than yesterday – 'old' experience is therefore less likely to demonstrate your full abilities,

(*b*) you are likely to be more familiar with current work, and therefore better able to discuss it in detail at interview,

(c)   documentary evidence is likely to be easier to assemble.

However, there is no requirement to use current experience exclusively – you may wish to draw on other experience to demonstrate your abilities more clearly. There is also absolutely no requirement to cover all aspects of your experience.

After these initial stages or coarse sieves, you now have to choose the ten Activities at which you are particularly adept (Level B – 40 points – 'do without supervision and able to supervise others'). One of the choice criteria at this stage will be the availability of documentary evidence, so it is perhaps better to have more than the minimum ten to select from, say twelve, in case one or more prove difficult to support.

Find and collect suitable documentary proof of your proficiency in at least ten Activities. Where possible, include Sector 0 'Professional and General' in the proof. Then assemble the evidence into suitably referenced packs (e.g. by Sector and Activity). Do not develop some complicated reference system of your own. The Reviewers are generally familiar with the numbering of the Sectors and Activities, so stick to that.

Writing the Report

This Report is totally dependent on, and inseparable from, the Activities you have chosen and the supporting documents you have collected to use (Chapter Nine):

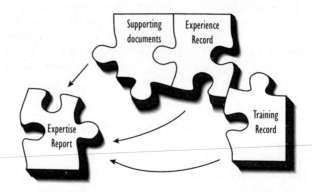

Only when the pack of supporting documents is reaching completion should you write a 2000-word Report on these Activities, explaining how you developed that proficiency and expanding on the problems for which the documents show the solutions. This Report could use only one project as the nucleus, but may draw on just as many as are absolutely necessary – no more. Do not be tempted to try to cover your *whole* experience – you will spread your material too thinly and open yourself to questioning on matters with which you may no longer be familiar.

Do note that this is *totally different* from the 2000-word Experience Report which those attempting CPR are required to write. If anything, it is the equivalent of their 4000-word Project Report – material to prove that you *are* what you are trying to prove you are, that you have developed and are about to prove the qualities listed in paragraph 4.4.

Finally, relate the paragraphs of your 2000-word Report to the Sectors and Activities and to the 'evidence' by a simple form of referencing; by far the simplest is to number the evidence envelopes as the Activities to which they relate, i.e. 4.5 being Activity 5 from Sector 4, and the documents within each as 4.5.1, 4.5.2 etc. If a piece of evidence fits more than one Activity or, conversely, one Activity is covered by more than one piece of evidence, then the referencing needs to be particularly clear and some kind of diagrammatic matrix may be necessary, showing pictorially how all the Activities and pieces of evidence interrelate.

The Activities can either be referenced (perhaps in bold type) within the text as they first appear, or in the right-hand margin of the Report. You may choose to write about each Activity under a discrete sub-heading, but this is not the only way of doing it and can be difficult for the reader to follow if not done carefully.

Is there a set format for this Report?

There is no set format for the Report and I have seen several different ways of setting it out. The obvious method is to write ten short paragraphs, one on each Activity, averaging 200 words each. But it is also possible, particularly where only one or two projects are being used, to write a narrative style Report which draws out each Activity, not necessarily in numerical order. This latter style is perhaps easier to read and therefore more user-friendly, providing the referencing is good. If you are having difficulty with the referencing, then the more formal format is possibly the one to use. Only you can decide, but remember that your decision will create an impression, so bear in mind the fundamental purpose – to prove that you are a professional engineer.

Your 2000-word Report does not cover your whole experience; the submission therefore includes a further 1000-word Report to enable you to describe briefly all your relevant experience (see Chapter Six) and describe how you have fulfilled the requirements of Sector 0 if you have been unable to cover them in this Report. Yet again, this emphasises my point that you cannot write the Reports one at a time; the submission must be compiled as a whole.

Chapter Eight

The Reports – common faults

Underplaying your hand

Nearly every submission Report I have read underplays the writer's hand; there seems to be a reluctance to spell out exactly what you did and understood, expecting the Reviewers to 'read between the lines' to decide for themselves. This is not good enough – you must *demonstrate* your abilities, not assume that they can be tacitly inferred. It is not generally in an engineer's character to boast, but in this instance, you must try! It is also difficult to admit to mistakes, yet these are often where good experience was gained. The Reviewers will not be so interested in the mistake as in what you did to rectify it and what you learned from it.

It took me some time to unravel why most candidates underplay their hand, but I think I know one fundamental reason. Engineers are problem solvers; once a problem is resolved and a solution found, they go on either to solve the problems of implementation or to another problem. In other words, they forget the original problem. So to write the Report successfully requires you to remember the solution and then unravel the thinking behind those decisions to expose the original problem.

Another difficulty is that we all know that, in reality, none of us works in isolation. Everything we do, we discuss with others. As a result, we are loathe to take personal credit for our work. But ask yourself who would have taken the blame if things had gone wrong — if it was you, then you were personally responsible, no matter how many people you discussed the problem with!

Every time you mention that something was done or happened, immediately ask the questions 'Why?' and 'What else was considered and why were the alternatives rejected?' as well as being sure to explain your precise role in the process. Be prepared to put yourself on the line and tell the Reviewers that, with hindsight, you now believe that there might have been a better way (after all, one of the things you must demonstrate is the ability to learn from experience). Even where the decisions were not yours to take, you can state that you offered advice, collected information, thought of ideas or suggestions or made recommendations which were subsequently accepted. You may even have drafted the Instruction, letter, Report or similar documentation for someone else to sign — say so! And where you had no direct involvement, you must show that you understand how the decision was reached; after all, it will not be long before you are making comparable decisions as a professional engineer; it is always better to have some experience *before* decisions need to be made than afterwards!

An example:

'The bridge was designed as a continuous, twin-celled cast in-situ reinforced concrete deck supported on piers with piled foundations and two spread footing abutments. The contractor proposed an alternative single-celled design in his tender which the Client accepted'.

What a throwaway! And perhaps a suggestion that his original design had not really been thought through? The candidate must know why the original format was chosen, what alternatives were rejected and whether one of them was a single-cell design. Why was the decision changed when the contractor offered the alternative? The candidate may not even agree with the decisions, but must demonstrate understanding of them.

I am often told by candidates that the reason why such explanations are not included is because they have been advised to leave questions hanging for the Reviewers to ask. I strongly advise that such a deliberate approach is mistaken for two reasons:

(a) the prescribed length of the Reports is too short to include everything, so there will inevitably be unanswered questions without any deliberate attempt to pose them,

(b) the Reviewers are wily enough to recognise a 'trailing coat' when they see one and will probably avoid it. There will be plenty of other questions which are not flagged up.

Another example of a case where the candidate hopelessly underplayed their hand:

> 'The piled foundations for each of the piers consisted of driven vertical and raked steel H piles founding at varying depths across the valley. In addition to the seven permanent piers, eight temporary piers were erected in between the permanent piers, again supported on steel H piles. Thus I gained much experience monitoring piling operations'.

Alongside the last sentence, a Reviewer had written 'What?'. When I discussed this experience with the candidate, I found out that he had in fact, devised a neat and simple device to aid the checking of the required sets on so many piles, thus saving hours of laborious and repetitive work. Yet he made no mention of this in the Report! His last sentence was wasted and the description could have been substantially reduced by a little longitudinal sketch to allow him space to expand on his ingenuity.

Vague generalisation

By far the most common mistakes relate to the manner in which you write. I can almost guarantee that you will repeat the most common of all – stating 'I was involved in . . .' or 'I was responsible for . . .'. To avoid these vague generalisations, *always* turn the sentences around to force you into being more specific – 'My involvement included . . .' and 'My responsibilities were . . .'.

Other very common understatements include phrases such as 'I was then transferred to . . .' or 'Once I had been given the task of . . .'. How much more positive it sounds and how much more in control you appear to be, if these statements are rephrased as 'I then transferred to . . .', 'I had the task of . . .' or even better 'I did . . .'. And the changes actually save words!

Writing in the third person

Many engineers have written reports for committees or clients, or for an academic forum, where writing in the third person is invariably required. There is frequently a tendency to write these submission Reports in a similar manner as a result of such experience. The worst scenario is where the Report actually gives the impression of being copied from a textbook:

> 'Control of a project is a continuous process which may be represented as shown. As a job proceeds performance should be monitored against the targets defined by the plan. Monitoring itself does not constitute control; this is exercised by making decisions on actual performance and updating the plan'.

But the most usual third person style is exemplified by:

> 'Many questions became apparent. How long would the design take? What resources would be required? What information was needed and at what stage?'.

Remember the underlying purpose! Rewrite as 'I identified the available resources and hence decided on a realistic time-scale within which I could programme the receipt of key information'. The length of the sentence has been decreased but, much more importantly, your responsibility is now *demonstrated*.

Use of abbreviations

Use acronyms whenever you can to keep the Report short, but always tell the reader what the letters mean the first time they are mentioned. In a previous example, in Chapter Six, I refer to an ARE, but what exactly is that? You might expect everyone to know, but you cannot assume that they do! There is another important point to be made here; since ARE is not a formal description of any person under any form of contract, the Reviewers have no idea of the responsibilities or role the candidate is playing unless you tell them!

Be careful that your acronyms are correct. I hate to think of the number of times I have seen r.c. concrete referred to — I am still trying to find out what it is!

Jargon

Avoid jargon! Obviously we all use it every day as part of our communication at work, but in formal documentary communication it is unacceptable. Here is a classic example, where I do not think there is one genuine engineering term in the whole sentence:

'The rebar cage was prefabbed outside the hole and craned in just before the RE's inspection so that the rubbish could be removed easily from the shutters'.

Such use of jargon is completely unacceptable. But worse – there is an implication here that perhaps the cleaning would not have been done at all but for the RE's inspection. So it is absolutely vital that you read and comprehend exactly what your sentences say, not what you think or believe they say! This version is much better, using fewer, more appropriate words:

'The reinforcement was prefabricated alongside the excavation and lifted in after the formwork had been thoroughly cleaned and prepared'.

The final stages

Even when you have edited the drafts down to approximately the correct length and made sure that every sentence is aimed at the objective of proving that you are a professional engineer, the task is by no means complete. Now you have to check the spelling, punctuation and syntax. If you have difficulties with these, do not tempt fate by being too adventurous. Like Winston Churchill, who admitted to early difficulty with the English language, keep your sentences short and your words simple, only use full stops and commas and master the use of the apostrophe.

It is very difficult to read your own work and find such errors, because you tend to read what you want it to say, rather than what it actually says. You also know what you are trying to say, whereas someone reading it for the first time does not. So you have to try to detach yourself and read the Reports as a third person. To help, it is also a good idea to get others, engineers who do not know anything about your experience and even people who are not engineers, to read your work. They will ask questions which cause you to query whether what you have written expresses exactly what you intended. For this book, one of the most helpful stages in the sequence of production was when the editor (who is not an engineer but does have a marvellous command of English) started to query what I had written; it really did make me think afresh.

Checking spelling is something else which requires great care and time; it needs to be done as a separate exercise, where you do not actually read the Report, but only check the spelling, punctuation and syntax. You cannot rely on the word processor's spell or grammar checking capabilities which are limited, generally Americanised and cannot tell the difference between 'right' and 'write', 'their' and 'there' etc. You need to be very painstaking — this example is a classic, which many people who read the Report failed to notice:

> 'This meant that both pupils and their parents were forced to cross a very busy dual carnageway'.

Do spread your checking beyond the Report itself, to the cover sheet and picture captions. It is not sensible to have a cover sheet which announces in two centimetre high letters that your submission is for the

'CHARTERED PROFESSIONAL REVUE'

Summary

By now I hope you will have realised that the compilation of these Reports is not easy. It is my considered view that there is a fundamental need for time in this process and I usually raise eyebrows when I suggest that at least four months is a reasonable minimum. There is no substitute for reflection — setting aside the Report for a few days and looking at it again with fresh eyes. You also need to ask as many others as you reasonably can to read your work — and not only other engineers. They will all offer different suggestions, but at least they will have caused you to look at the Report in different ways. You can then make your own mind up as to which suggestions you incorporate and which you discard.

Make the Report and supporting documents look authoritative. Bind them neatly, using an appropriate proprietary system. There is no longer any restriction on the binding method (there used to be a requirement to use treasury tags only) but do consider how the documents are going to be used. There is nothing more annoying than a document which closes every time you let go! Choose a form of binding which enables the pages to lie flat - slide-on plastic spines are entirely inappropriate and invariably seem to come apart and allow the pages to spill out. If using plastic ring or spiral binders, make sure they are of the correct size for the thickness of document; so many submissions arrive with outsize spirals, making stacking difficult and making the document look second rate.

Put a cover page on the front of each document, deciding first on a uniform style for the entire submission. Do not use one of the overly sophisticated publishing possibilities thrown up by computer technology — they will not impress the Reviewers because they only show that you (or a colleague) have apparently limitless time to experiment with the various programs. A clean, clear cover is the most impressive. Avoid scanning in your company logo or, worse, the Institution crest. The documents do not belong to either your company or the ICE, so you could well be breaking the law if you do! Think about what is important — to you this may well be that the documents comprise a submission for the professional review but, to your Reviewers, the most important information is who you are and which document represents which part of your submission. So state clearly on the cover sheet both your name and the Report's title, reducing information such as the type of review to a secondary role. Be proud of your work — take possession of the Reports and documents by signing them on the front cover rather than at the end.

Since, in theory, you have infinite time and infinite support and resources to prepare these Reports, the Institution expects them to be perfect, just as any professional report emanating from your office should be. Whenever I ask candidates after their Review if they have any advice for those candidates following them, their response is invariably 'I wish I had allowed myself more time!'.

Chapter Nine

Appropriate supporting documents

For the Incorporated Professional Review, candidates are required to submit 'documents which will assist the Reviewers to establish the level and standard of work' (paragraph 4.5iii) and 'are advised to keep suitable documents during their training/career' (paragraph C5). Such documents are to support the 2000-word Report on Activities only.

ICE 101 for the Chartered Professional Review is a little more specific:

'The reports should incorporate numerical *analyses, cost data* and *drawings* or other *illustrations* as appropriate to show adequate *understanding* of technical, financial and environmental implications of decisions'

(paragraph C3).

It seems to me that there is no fundamental difference between the requirements for supporting documents for either Review. Clearly there will be more evidence of breadth for a CPR candidate, while there must be strong evidence of the candidate's technical and professional competence and ability to exercise engineering judgement for IPR. The aim is to provide supporting evidence which assists in demonstrating the abilities being displayed in your Reports or which justifies and supports the decisions which you took. The interesting point is that ICE 101 states that both Reports can incorporate documents, but the situation is not so flexible in ICE 102. My view is similar to the method I used to employ in reading specifications; exploit any uncertainty to your advantage!

Careful choice is necessary; the Reviewers are not impressed by sheer bulk. The inclusion of every document must be a considered decision; does it contribute positively towards the overall objective – to prove you are a professional engineer? If it does not, then discard it.

The following advice is therefore applicable to both Reviews. It may well be that a specific IPR submission does not encompass every section which follows, but a great deal of specific guidance is given in the Schedule of Sectors and Activities about the sort of documents which might prove satisfactory for IPR.

Analyses

These must demonstrate that your decisions (or the decisions taken by your line management, but to which you contributed) are based on sound technical knowledge and understanding. Analysis is not design, but is an integral part of the design process and generally takes place towards the end, when you are justifying the solution you have chosen and making sure that it will withstand the forces and loading conditions you have decided it must resist. Analysis (generally using quick design methods) assists in the choice of the most appropriate solution and then satisfies you in detail that the chosen solution is adequate.

Documentary evidence is needed for the Reviewers to ratify Technical Competence if they feel it necessary, but they are more interested in how the calculations were used as an integral and vital part of solving a problem or implementing a solution. The purpose is to *demonstrate* technical understanding, not to show that you can 'do calculations'!

Do not rewrite calculations specifically for use in the submission; documents should have been prepared 'during the normal course of your work'. Do not abstract aborted calculations; they were an integral part of the design process, vital in choosing the most appropriate solution. Do show that you know how the loading conditions were chosen, even where they are established by others – be able to answer such questions as 'Why was this designed for a 1 in 25 year storm?' or 'Why did you consider earthquake loadings?' As long as your calculations are neat and legible (even a few crossings out are acceptable), you state where new figures or numbers came from and explain why you did them that way by annotating the calculations appropriately with cross-references, then the Reviewers will be able to see clearly that not only did you do the calculations, but that you understand their function.

The same criteria apply whether you used a computer program, did the calculations by hand or got someone else to do them.

- What assumptions were made or are inherent in the computer program? How did you satisfy yourself that those assumptions were valid for your problem?

- Was the method or program you used the most appropriate? What others were considered and discarded?

- How did you make your problem fit the computer program or method?

- How were the critical loading cases decided upon?

And, the most important of all:

- How did you satisfy yourself that the results you obtained were realistic?

These sorts of question make it quite clear that calculations are a means to an end rather than an end in themselves. You must *demonstrate* how the calculations were used to solve problems such as deciding on alternative structural forms, sizing and reinforcement of members, choice of materials, or the most appropriate construction methods and plant. For construction method statements, for example, you might have had to consider temporary stability.

Cost data

A conventional bill of quantities is perhaps the best vehicle to *demonstrate* an understanding of construction methods and the financial implications of the solution, but it should include something to demonstrate your understanding of rates and item coverage – the 'build-up'. It is important that you do more than use previous rates from other estimates or quotations so that you *demonstrate* your *understanding* of how the rates were adjusted to suit your particular job, or what a rate includes – Item Coverage. You would not have to be a professional engineer to copy previous rates into a new bill, but I am not at all sure the result would actually be realistic! The job for which the cost estimate is compiled need not necessarily be a construction project; it could, for example, be the manufacture of apparatus or equipment, or a traffic count.

Other possibilities might include:

- the substantiation of a claim, particularly where it involves operations not included in bill rates

- the estimate for a variation order

- the build-up of an estimate for a proposal

- an estimate of design costs for a client.

Drawings and illustrations

Use drawings throughout your submission to save words – 'one picture can tell a thousand words'. Drawings need not be contract drawings – in fact large drawings are not a good medium to use for the submission since they are extremely difficult to unfold and fold. It is preferable to include sketches within the text or the salient parts of a contract drawing (perhaps even simplified). Other kinds of illustration might include:

- photographs

- architect's impressions

- public explanatory leaflets.

Be careful in your choice of the first of these; it is surprising how often a photograph shows something in the background on which you would rather not be questioned!

You must also demonstrate that you can communicate by drawing and can visualise an engineering problem (spatial awareness), so include drawings and sketches (not necessarily exactly to scale or drawn with a straight-edge) where you thought out a design or construction problem or transmitted information – e.g. to CAD or to the subcontractor or designer, or to someone manufacturing apparatus. You *will* be expected to draw during your interview – paper is provided.

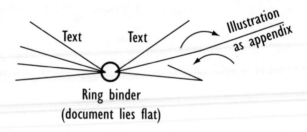

Ring binder
(document lies flat)

As far as possible, try to incorporate the supporting documents in such a way that the Reviewer can refer to them at the same time as continuing to read the Report. Obviously, if the support is a substitute for words, then it should be incorporated into the text whenever possible, but a bulky appendix giving the detailed justification for a technical or financial decision is better bound in at the back. But do think the problem through! If you are going to incorporate photographs into your text, then you are incurring the substantial additional cost of photocopying in colour. Can it be justified? Would it be better to paste the photographs into appropriate spaces afterwards or would this detract from the overall appearance? Could the pages with colour be photocopied and bound in separately? One very useful device for overcoming the problem of referring to appendix material while reading the Report is to print the appendix on the right-hand side of A3 paper (in a similar manner to the ICE Essay booklet), so that the information can be folded out alongside the text and then folded back when finished with (as diagram above).

Chapter Ten

From submission to review

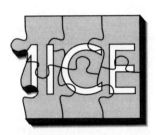

By now you have compiled six of the eight pieces of your jigsaw and the picture should be becoming clear. Now is the time to finally make sure that all these pieces interlock without either gaps or overlaps. Each document should stand alone as a complete entity, but be referenced to the others. Supporting documents should be rechecked to make certain that they really do *support* your case or provide the Reviewers with essential background information. It is at this stage that I get frantic phone calls:

'My submission weighs more than 1 kilogram!'

What these callers have done is to put their entire submission on the scales and frightened themselves because they have only scanned, rather than read and understood, the specification.

You are in fact going to send three parcels — one by the 15th of the submission month to the Institution which will weigh much less than 1 kg and, later, two more to your Reviewers. One of these must weigh less than 1 kg, the other will almost certainly weigh slightly in excess of 1 kg. Yet you will be complying with the rules! Reread them and make sure you comply with the spirit of them. ICE 101 is rather like the Highway Code — you cannot be prosecuted for breaking the code, but such behaviour can be used in evidence against you if you subsequently have an accident. Very few Reviewers to my knowledge, actually check the weight, unless the submission feels heavy (remember, they have to carry several around) or is full of superfluous material.

About a month after the closing date for entries, you will receive a letter from the Reviews Office informing you of the names and addresses of your two Reviewers. Send your packages off to them as soon after this as possible. The rules require receipt 'two clear weekends' before your interview, but it surely creates a good impression of organisation and control if you get your parcel to your Reviewers before the deadline.

Make sure your parcels arrive — and in one piece! Since you cannot phone your Reviewers to check, send the parcels by some form of recorded delivery so that another uncertainty is removed — you will know they have arrived safely. Package them securely in a purpose-made envelope (available at all stationers and the Post Office); this is so much better than constructing a parcel. Reviewers like to use the envelopes as a filing system, so it also helps if you put your name and candidate number on the back flap with a felt-tip pen.

Make absolutely certain that you use the Reviewers' names and qualifications exactly as you were given them. There is nothing more annoying than receiving an incorrectly addressed letter and, because of its familiarity, it is something which the recipient notices immediately. Take the Reviews Office letter with you, so that if a Reviewer does complain, you can show them precisely what information you were given.

Would your organisation send any documents to a client without a covering letter? Yet many submissions arrive cold and unannounced. Write to your Reviewers, if only to tell them how much you are looking forward to meeting them! Confirm the date and time; this way you avoid another possible uncertainty.

The waiting period

It is normally nearly three months from the final date of submission to your interview. This is the time to look ahead, to try to anticipate how the remaining two pieces might be compiled; you are not totally in control of these but, as far as possible, you must reduce the chances of the unexpected. For example, you should be able, perhaps with help from other people in the office, to make a realistic guess as to what topics might be covered in the first essay or written test, since both are based on your own personal experience.

Should I research the Reviewers?

It is always a temptation to try to find out who your Reviewers are and what their particular field of expertise might be. This is, perhaps, an understandable effort to remove yet another uncertainty, but I fear it is misplaced and can, on occasion, backfire. You will not get any information from the Institution. I, and my colleagues in the regions, receive infrequent phone calls asking us about Reviewers; we are also noncommittal in our responses.

The reasoning is straightforward. Your Reviewers have been chosen for their expertise *as Reviewers* and not particularly because they are able, from their own particular field of work, to ask searching technical questions. Remember the purpose of the Review — to explore whether you have become a professional engineer, not whether you can technically do your job.

All the Reviewers are volunteers, who serve the Institution in their own time and largely at their own expense, apart from travelling and accommodation. They carry out this service for two reasons: one is their professional responsibility and a desire to serve their Institution in a tangible way, the other is that they genuinely enjoy meeting 'the next generation' and really look forward to successful reviews. Most were trained themselves some 25 years ago and all have an interest in training and development and the maintenance of the Institution's reputation for excellence. The Institution is very proud and protective of its Reviewers, who are themselves subject to continual monitoring and review. Most are spotted as potential Reviewers by the Regional Liaison Officers and asked to volunteer their time, usually because of their involvement with younger engineers' training. Their initial response is almost invariably a doubtful, 'Do you think I am good enough?' All recent recruits go through quite a lengthy process of training, at any stage of which the Institution may, without prejudice, decide not to continue.

The difficulties in researching your particular Reviewers come when you perhaps find out that one has a reputation (probably totally false because you cannot possibly sample a representative group of their candidates) for a particular line of questioning and you become obsessed with that to the detriment of other aspects. The interview is almost certainly going to be perversely different! Or you might hear that a Reviewer is particularly aggressive (again, probably without any real justification) – how is that going to affect your build-up? Such a report is hardly likely to give you confidence. Having observed for several years, I know that most Reviewers do, in practice, change their style slightly to suit each particular candidate.

The final danger of research is that you can easily be misled. One Reviewer has spent his entire career in the South and Midlands. A candidate found this out and felt very secure describing and drawing a problem in the North; he was disconcerted (to say the least) when he realised that the Reviewer knew the area well – because relations live there.

Of far greater significance is deciding how to make the presentation and what it should cover.

Preparing for a presentation

I am well aware that the current ICE 102 *Routes to Membership – Incorporated Engineer* does not state that a presentation is going to be required; but many Reviewers start the interview by asking candidates to 'take us through your submission' – disconcerting if you have not anticipated such a start! This presentation is the key element locking the submission to the Review itself and is a great opportunity to influence the whole process; it will occupy approximately a quarter of the interview and could well determine the shape, content and style of the remainder. It is the first opportunity for you to actually show yourself operating as a professional engineer rather than writing about yourself.

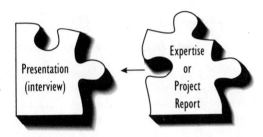

Why do the Reviewers want a presentation?

After all, they have already an awful lot of information from your submission! The formal introduction of a presentation fairly recently really only ratified an approach by the Reviewers which had been apparent for some time – to start by asking, 'Would you like to tell us a little about your project?' This approach reinforces the view that the interview is *your* chance to strengthen your case and the Reviewers expect you to do most of the talking, with them only directing you to those matters they would like you to cover.

The presentation has been very effective for the Reviewers at both Chartered and Incorporated Reviews, so do not be surprised if your IPR starts in the same way – indeed, a presentation may well be included in a revised ICE 102.

This part of the Review is meant to replicate the sort of presentation you will be expected to make as a professional engineer when putting proposals to a client. You will have sent them documents beforehand, but would follow these up with a visit to draw their attention to the key issues and why you believe your organisation is the best one to undertake the work. So in this case the purpose of your presentation is to draw the Reviewers' attention to those aspects of the project where you displayed your full ability and convince them that you are a professional engineer.

It is not a good idea to sit and read your Project Report verbatim, although it is just about possible if you hurry. It has been tried –

unsuccessfully. You do need to quickly sketch in the problems and then move on to pick out the key points in the work which *demonstrate* your skills and abilities and reinforce them.

What visual aids can I use for my presentation?

Almost any prop can be used in the presentation – as long as it is appropriate and makes a positive contribution to the fundamental aim of the Review. Remember, you are going to talk to two people across a small table. I do strongly recommend you to use something – anything which will prevent the Reviewers spending the whole fifteen minutes staring at you, which would be very disconcerting.

The classic aid is an A3 flip chart, with pictures and diagrams on the Reviewers' side and bullet points on the back. This is a convenient way of keeping yourself on track, both with content and timing, but most of the presentations I have seen have used far too much material – too many photographs on one display, too many page displays. After all, you are expecting the audience to assimilate the information virtually instantly as an adjunct to what you are telling them, so they cannot realistically be expected to take in much. I fear that many visual aids are compiled to assist the presenter, rather than the Reviewers. When being coached as a lecturer, I was told, 'Never more than four lines, never more than three words to a line'. A maximum of twelve words on a display – not a great deal of scope. But if you look at advertising hoardings, they do not seem to stray far from that advice. For a 15 minute presentation, more than one page every three minutes is likely to be excessive, so aim for a maximum of five pages.

In this scenario, I would suggest that a movie projector or video with screen are both inappropriate, although they have been tried. Neither are contract drawings much use; unfolding them does not cause too many problems, but folding them back up certainly does! Candidates do tend to look as though they are trying to wallpaper the room.

I learned recently that desktop computers do not work either (even when battery operated). Apparently, modern display screens are designed to be very directional, so that the person sitting next to you on the train or plane cannot see what you are doing. It is therefore almost impossible to adjust the screen to be seen by the two Reviewers. Anyway, how do they know that you set up the material yourself? Your organisation's computer wizard could have done it all for you. Do not use them! Similar remarks apply to battery operated slide projectors. All are inappropriate.

More appropriate would be something to put on the desk. What that should be is best left to your imagination and creativity. I have seen all kinds of props — site plan jigsaws (where a development progressed through a series of overlays), scale models, pieces of rusty iron, a disc cut from a suspension cable, a piece of apparatus. The important thing is to make sure by adequate practice that you can effectively handle any props, even when very nervous. Fumbling about in an attempt to make something work will not impress the Reviewers.

This brings me to another point. You will, I hope, take documents with you in case specific questions are asked about matters which did not have supporting documents in the submission. If you need to refer to them, do make sure they are catalogued in some suitable order in your briefcase and that you have practised finding them. You do not want to spend great lengths of time bent double behind the desk, with the Reviewers wondering what you are doing. Again, take 'just enough'; be ruthless and discard anything which is not going to reinforce your case significantly — not 'just in case' but vital.

Do I need to practise?

Having drafted a script, you must then practise. Even famous and apparently effortless public speakers invariably practise. I know one such speaker who *never* allows himself to be persuaded to make a speech without having prepared something first, perhaps not specifically for that occasion, but after many years' experience, he has sufficient material and can adapt it to a rough outline in his mind. He also regularly stands back from a full length mirror and critically observes how he uses gestures and movements to reinforce his message. Yet most people believed he was a brilliant impromptu speaker who found it easy to talk at any time on anything. I do not think that anyone ever really finds public speaking easy, however long they have been doing it. But done well, it can be very satisfying.

The ideal place for you to practise is at a dressing table (since you will be sitting down during the interview), where your image in the mirror is about as far away as the Reviewers will be. Look yourself in the eye, watch how you move and, particularly, see how you handle your visual aids. Speak out loud, pitch your voice at your reflection and time yourself (you will feel very self-conscious, but this is an advantage because it actually causes the adrenalin to flow, replicating the anxiety you will feel on the day – your timing is therefore likely to be about right). Timing a talk cold or in your head is never successful; everybody goes rather ponderously, whereas in the real situation you will either put in extra material and go even more slowly, or you will gabble and finish early! Which scenario applies to you?

If you intend to use some form of prompt, keep checking that the words on it are exactly what you need; each time you are disconcerted during practice, see whether better keywords or a different layout would help, until your prompt is honed to perfection. Personally, I believe that if you practise hard, you will have no need of a comprehensive prompt; perhaps just a list of the key points. This leaves your hands free to manipulate any visual aids.

Having done all this, get colleagues, relations, indeed anyone to listen to you. An ideal opportunity is the Graduates' and Students' Papers Competition, even though the scenario is different you will stand and use bigger visual aids for a larger audience. Seek opinions, criticism and advice you do not need to take it all, but it will all help. Furthermore, you will become more confident and begin to relax a little.

Nevertheless, on the day, you *will* be nervous – and this will work to your advantage! Any actor will tell you that you have to be nervous to give a good performance; the day they become complacent is the day they leave the cast. What you must be able to do is control and conceal your nerves and the only way to accomplish this is to practise. Actors learn techniques for controlling their nerves — deep breathing, shaking and relaxing their limbs; there are all sorts of techniques which help. Throughout your previous training, you should have been taking every opportunity to speak in public so that you have developed these skills. Few of you will actually have done so; you need to practise and learn now.

The other aspect which probably needs attention is the written work you face in the afternoon and I devote Chapters Twelve and Thirteen to considering this whole aspect, since it is a major cause of difficulties for so many candidates.

Chapter Eleven

The Review day

Presentation and interview

What happens on the day?

The straightforward and obvious answer is – 'Who knows'. While the Reviewers will by then have a broad plan of what they wish to cover and will have decided on the titles for the afternoon's written work, even they will to some extent play it by ear. Nevertheless, there are many uncertainties which can, and ought to, be removed. Any uncertainty breeds doubt and doubt breeds nerves, so the fewer uncertainties there are on the day, the better you are likely to perform.

The first uncertainty is the journey – how and when you are going to get to the venue you chose on your submission form. If you are travelling the night before, have you arranged somewhere to stay? If at all possible, my advice would be to stay with a friend, either in their home or together in bed-and-breakfast close to the Review venue (the full rate in most of the provincial venues used by the Institution and those hotels local to the headquarters in London is usually exorbitant). Being on your own in a strange place late in the evening with no one to talk to can be daunting and confidence sapping; it is not the best preparation for a full, stressful and taxing day.

Whatever arrangement you decide upon, it is sensible to do a reconnaissance. If you intend to travel on the morning of the

Review, then do the same journey at roughly the same time on a weekday to gauge the traffic or the time on public transport. When stationary in a jam, it is comforting to know just how much longer it will take to arrive when you finally start to move again.

Visit the venue, get the feel of the place. The Institution in Great George Street is in fact quite a friendly building, but you will probably find it very daunting if you visit for the first time on the Review day. Have a drink in the bar, either there or in the provincial hotel venue. Find out what you can do (gym, swimming pool, sit in the park, etc.) between the interview and the written work. This period could be as long as three and a half hours, during which time you must be able to relax as far as possible; sitting in your car or in a café revising merely makes you more tired and seems inevitably to remind you of all the things you do not know — definitely not the best way to prepare for the afternoon.

What should I wear?

The answer is as it has been for every aspect of the preparation — what are you trying to demonstrate? That you are a professional engineer: so look like one! My only criterion is that you should feel comfortable and enjoy whatever you are wearing; you do not want another uncertainty on the day. Do not wear something which you continually have to tug or adjust because you feel uncomfortable. Do not buy a dark formal suit if your style is somewhat more adventurous, but at the same time remember that most Reviewers are probably nearer the age of your parents rather than your age — what would they expect? Most Reviewers, I think, would expect a jacket but not necessarily a suit, with a tie for men and modest neckline for women; few engineers (particularly the older ones) like or empathise with power dressing or strong fashion statements. Again, it is a question of judgement and the impression you are creating.

I have also been asked (by a man) what the Institution's attitude is to long hair. My answer was, 'We think it's lovely!' from which he took me to mean that he should get it cut. No! I asked him whether he ever represented his organisation to the public and whether they took him and his views seriously. The answer to both was a resounding yes, so he did not need to cut his hair, which in any case was extremely well groomed. Perhaps he did put himself initially at a slight disadvantage with more staid Reviewers – if that was the case, he overcame any resistance and was successful.

The weather on the day may mean that you have to wear a coat or carry an umbrella. Leave them in the cloakroom to avoid the clutter which would otherwise result as you enter the interview room.

What happens on arrival?

There will be a lot of activity in the building on the day. There could be as many as twenty Reviews all going on at the same time, so there will be upwards of sixty people directly involved in your 'round' and more from the one which is just ending. Find the reception desk and book in as soon as you can, before finding the cloakroom (you should know where it is beforehand) and depositing everything you do not need. You will probably already have a briefcase and possibly some form of presentation material, so do not clutter yourself further with a raincoat and umbrella. Remember that the first thing that one of your Reviewers will do, after coming out to the reception area to invite you to join them, is to shake your hand – keep your right hand free! Otherwise the very first visual contact you have with your Reviewer will put you on the defensive, as you fiddle about getting flustered, trying to pick everything up and then changing everything around to shake hands. This is why I recommended in the paragraph above that you relieved yourself of any unnecessary coats and umbrellas. And this reminds me of another vital point; nerves make us all want to go to the toilet, so make sure you go before, and do not feel the need to during, the Review! You do not need any unnecessary distractions.

From that first moment of contact you are being observed! At least look as though you mean business, even if you feel awful. It may interest you to know that the Reviewers have usually only just met each other, so they are nearly as unfamiliar with each other as they are with you. Hopefully you have remembered both of their names (which you were given for posting your documents) so even in the heat of the moment, you should not forget who they are. Sometime during the interview, refer to each of them by name – it suggests you are in control of yourself and the situation.

During the short walk to the meeting desk, your Reviewer will talk informally, probably telling you a little of their own background and asking whether you have had a reasonable journey – anything to try to get you to relax. Talk to them! Do not simply reply, 'Yes, thank you' but have something in mind – 'Yes, but did you get stuck in that sewer replacement job in the High Street?'. Such an answer shows that even when under pressure you are interested in what is going on around you and had the foresight to anticipate possible delays; even casual conversation is creating an impression and commencing to develop a relationship which must blossom and reach fruition in the next hour.

The interview

Your desk will be one of several, arranged round a large room in such a way that, as far as possible, candidates are not in eye contact. It sometimes happens that one of your senior staff, or someone you know, is acting as a Reviewer on the same day; if you know that this is a possibility, do inform the Institution and strenuous efforts will be made to ensure that you are not in eye contact with them either.

Your two Reviewers will sit opposite you at an average sized table; there is not a lot of room after you add two or three drinks and two sets of submission documents! There may be a third person sitting at one end of the desk, within your eye range but taking no part in the proceedings. This could be an observer from the Institution or the Engineering Council as part of the quality assurance processes, or a new Reviewer learning by shadowing.

If no drink has been provided for you, ask if you may go and get one – there is usually a table somewhere in the middle of the room supplied with water and other soft drinks. Nerves will make your mouth dry, so be prepared – always look in control.

How should I make the presentation?

While the rules may not state that you have to make a presentation, it is common for the Reviewers to start by asking informally for you to 'take them through' your Expertise or Project Report. So be prepared, whether it is a stated requirement or not.

Picture the scenario when deciding on how you will make your presentation and particularly when you practise, as you must if you are going to reduce all the uncertainties which otherwise surround this aspect of the day. A projector and screen or video are not deemed appropriate for this intimate setting; you do not need a pointer or light pen and anything you use to illustrate your work should be to the scale of the setting – A3 not A0! After all, there is an audience of two, not two hundred. Make sure, if necessary by asking, that both Reviewers can see your visual aids; so often a candidate puts a flip chart on the left corner of the desk, forcing the right-hand Reviewer to lean across to be able to see properly. Move your seat so that your visual aids are centre stage during this formal part, but remember to move it back as you finish and you become the centre of interest.

I personally would not stand throughout, although I have seen candidates become so enthralled in aspects of their work that they have momentarily stood up to demonstrate a point. And this really is the key to success. Even though you will have practised many times, you must not give the impression of going through the motions; your Reviewers will be genuinely interested, so do not dull their interest by boring them. You must inspire them with your enthusiasm.

At the end of your presentation, give your Reviewers a clue that you have come to the end; they may otherwise think that you are merely pausing to gather breath or have momentarily lost the thread. Make sure that you stay just within the time limit for a formal presentation and certainly no more than ten minutes if you have responded to an invitation to give an informal presentation; anything longer will start to irritate the Reviewers, who are aware of how much ground remains to be covered in such a limited time. You are unlikely to get away with the tactic of extending your talk to reduce the questioning.

What kind of questions will I be asked?

The questioning which follows the presentation is really a series of prompts to set you off talking again in a direction determined by the Reviewers, so do not give monosyllabic answers. They wish to see an engineer's mind at work, talking through problems and difficulties and discussing the judgements which have to be made every day of our working lives. Do not be disconcerted if you are asked questions you cannot answer. I have discussed this with Reviewers, who have told me that occasionally they do not know the answer either. What they want to see is what comes into your mind, how you tackle the question, what you would need to find out to arrive at an answer – in other words, what you do every time you are faced with an apparently insoluble problem at work. As one Reviewer described it, 'Engineers stumble towards a solution in a relatively organised way! I would like to see them doing just that'. This is why I counsel that the Review is very different from the conventional examinations you have experienced up to now, where you have been expected to give authoritative factual answers to any questions you have been asked. Engineering problems can not be resolved by such black and white decisions. The Reviewers are anxious to see that you are capable of *judgement*.

If you feel yourself coming under pressure, it is because you are not delivering what the Reviewers expect. Try to detach yourself, become a third party and mentally step aside to see what it is they really want. After all, you have convinced yourself and your sponsors that you are capable and competent, so you ought to feel confident that you are able to persuade these two fellow (or peer) engineers.

One of the things which really surprised me when I first became involved in the administration of the Reviews, was how much laughter emanates from the room, particularly towards the end of each batch of interviews. Looking back to my own interview, for which I was extraordinarily well prepared by my Chief Engineer, I recollect that we had quite a jovial discussion on one particular aspect of my experience not mentioned until the very end – it turned out to be the essay topic. The Reviewers really do enjoy meeting good candidates; a good morning for them would consist only of passes. They do not enjoy interviews where progress has been 'like drawing teeth' or meeting someone who was inadequately prepared or not up to standard. You would probably not enjoy that sort of interview either, so do not tempt fate.

One last point about the morning – your two Reviewers will be delighted to relieve themselves of their copies of your submission, so you will be coming out of the room with three sets. It is always rather nice to be able at this stage to draw out a plastic bag and put their documents into it; it shows you had anticipated that they would hand them back and that everything is under control. Beware the quick throwaway question just as you relax – you are still under observation until you leave the room.

Waiting for the afternoon's written work

Try not to draw any conclusions from the way you felt the interview went. First, you will not be able to remember accurately – it will quickly have become a blur! Secondly, the Reviewers are unlikely to have revealed their thoughts and opinions, either about you, your ideas or your experience; so any conclusions you come to are based on very little factual evidence. I have known candidates to burst into tears or become extremely angry, convinced that they have failed, when in fact they did really well. Others have come out feeling very confident, only to get a letter in due course pointing out where they went horribly wrong.

You may, like a boxer, feel battered and bruised; just try to return for the next round – the written part – convinced that you are winning! This is where you need your friend again; like a good second, they need to bolster you up for this final round. There is no advantage in shutting yourself away in your car and mulling over everything that went wrong. This is what you inevitably will do; even though I have been lecturing now for nearly ten years, by the time I arrive home after a long, lonely drive, I have convinced myself that 'Tonight it was a disaster!' I do not know why the human mind works in this way, but it does seem to for most of us. It is far better to join other candidates and talk, not about the interviews but about your work and experiences, last night's match, the latest music craze – anything to take your mind away from the morning. It all helps you to relax in preparation for the last burst of adrenalin; like any good athlete, you must pace yourself.

Chapter Twelve

The Essays and Written Test

The Purpose of the essays and written test

This final part of the Review brings the whole jigsaw to completion, locking all the pieces together. It is often said that this part is the most common cause of failure but, analysing the figures in rather more depth, it quickly becomes clear that if the written work is removed from the statistics, the Review pass rate hardly alters. This means that those whose written work is inadequate are generally also found wanting in other aspects, or pieces, of the jigsaw. The implication is therefore that all the main pieces of the jigsaw will become connected to this final piece – all your experience, your expertise from the Project Report or supporting

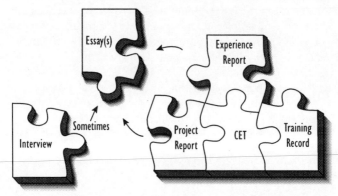

documents, your CET and your training. I have even known instances where the Reviewers felt that a candidate did not give as good an answer as they were capable of during the interview and so asked the same question again as an essay:

What is an essay?

This is a fundamental question which many unsuccessful candidates appear never to have addressed. Most have not even looked the word up in a good dictionary, where they would have found that an essay is 'an effort to do something; attempt; trial'; 'a literary composition, analytical or interpretative, dealing with its subject from a more or less limited standpoint'; 'a composition, shorter than a treatise, on any subject'.

The origin of the word is generally said to be 'esei' – to attempt, to try. Some prefer 'esai' – to purify, to assay, distillation (as of metals). Both seem appropriate in this context.

From these definitions it is evident that an essay is an attempt at concise literary composition. This accords with the Institution's requirements of 'a reasonable first draft' and the emphasis on 'Clarity and Presentation' and 'Grammar and Syntax'.

The essay format is not one normally used by civil engineers, whose literary prowess is not often tested; we write many things – reports, theses, dissertations – but not essays. We need to look for guidance and advice in literature. The collective opinion can be summarised as 'a literary composition devoted to the presentation of the writer's own ideas on a topic and generally addressing a particular aspect of the subject. Usually but by no means invariably, brief in scope and informal in style, the essay uses an elegant and concise non-fiction form to convey and establish opinions, without obstructing the writer's mind or character. It therefore differs from such formal expository forms as the thesis, dissertation or treatise. Incisive and thought provoking'.

In my search for guidance to help potential candidates, I eventually found what I consider to be the most useful definition of them all – by Francis Bacon:

'carry on your discussions with clarity and power and rigorousness, in recognisable sequences of enquiry, discovery, expansion, challenges and conclusion, all conducted with reason and addressed to any subject that takes your fancy'.

Francis Bacon was one of the first Britons to relate scientific discovery to practical problems through meticulous research and progressive experiment. It is perhaps not an unusual coincidence that, as an early exponent of what we would now describe as engineering thought, he should also have produced a description of an essay which I believe is of inestimable value to civil engineers.

Why does the Institution expect me to write essays?

An essay is a non-fiction exercise in prose which expresses, in elegant and precise form, the considered opinions of the writer on any fact, subject or concept, without obstructing the nature of the writer's mind or character.

Why does the ICE want Chartered Engineers to use this literary format which is unfamiliar to them? Why is the comparable test for an Incorporated Engineer called the Written Test? The answers lie in the fact that there are several differing forms of essay question, each of which requires a different style of answer.

Many people have tried to define essays, but an equal number of writers has continually redefined the boundaries in the name of literary freedom. In the context of the professional reviews, I believe that it is reasonable to divide the essay into four broad types:

- *Factual* – e.g. 'Outline the stages in the development of a design for a (bridge/dam/leisure centre . . .) from the initial brief to the issue of working drawings, giving examples from your own experience'.

- *Expository* – e.g. 'Discuss the records which should be kept by the Resident Engineer responsible for the construction of a sewage works. Describe the content and function of each item'.

- *Argumentative* – e.g. 'Discuss the advantages and disadvantages of using concrete additives in the construction of motorway pavements'.

- *Visionary* – e.g. 'Engineers are born not made. Discuss'.

In general, although not invariably, the first of the two Essays or the Written Test follows one of the first two formats above, where you are expected to be able to communicate in writing facts and procedures with which you should be familiar, since the questions are based on your direct experience. This type of essay should enable you to demonstrate:

(a) the ability to marshal facts quickly into a reasoned order to a tight deadline,

(b) that you are decisive, clear thinking and can 'get it right first time',

(c) that you are able to explain something with which you are familiar to someone who is not,

(d) that you have an incisive ability to 'see the wood from the trees', without becoming immersed in repetitive or irrelevant detail.

These are the characteristics of any good professional engineer.

The latter two styles ask you to state and justify your own views and opinions and represent the questions set in the second Essay for CPR. There is no 'correct' answer to such questions, indeed the Reviewers are expressly told that they cannot mark an essay down merely because the views expressed differ from their own, with one very important proviso – that the opinion is not mere prejudice, but is one which could reasonably be developed from the experience of the writer. These latter types enable you to demonstrate all the above qualities, with the addition of:

(e) independence of thought and opinion,

(f) a professional attitude, integrity and honesty,

(g) judgement for the benefit of the general public good,

(h) wide knowledge and understanding of current affairs

all attributes of a leader of the profession – a Chartered Engineer.

These scripts are not in any way like the written examinations taken during an academic course; they are not a test of knowledge, but a test of your abilities as a professional engineer, in which knowledge is used to *support* your responsible reasoning and arguments (hence the marking of 'relevance' — of your knowledge to the question and to your opinions).

How should I set about the Essays?

Consider the keywords which come out of Bacon's definition, coupled with the notes in the ICE booklets, and compile them into the framework or pattern:

Enquiry — what *exactly* does the set question ask? You must read precisely what it says, not cursorily assume it is the question you hoped it might be or think it is. What are the keywords? Dissect the question by underlining them. Look at each in turn and decide what it means. Why has this question been asked? What is the relevance of it to your experience?

Discovery — having identified which parts of your experience the question relates to, you should now be able to list your relevant knowledge and then add any further background reading or facts which you have.

Expansion — where does the question lead? From your assembled knowledge and experience you should now be able to develop a discussion or a line of reasoning.

Challenge — as part of your extrapolation you may now be able to provoke thought by debating your own understanding and opinions, supported at each step by relevant experience or knowledge.

Conclusion — *your* personal opinions or thoughts as a summary of what you have already said in detail.

In effect, an essay leads the reader from the question, through a reasoned and logical discussion or explanation which they should follow and accept, to a conclusion which sets out your views on or understanding of, the question, with which (hopefully) the reader will then agree, or which will provoke them into thinking more about it themselves. Newspaper editorials are a useful source of ideas on approach; note how, while the content differs from day to day, the format or technique is nearly always the same each day. You will have heard the trite description, 'Tell them what you are going to tell them, tell them, then tell them what you have told them'. I do not think this approach is relevant or advisable in this situation, where time will not allow much repetition. You must, however, have the time to summarise and conclude your work.

Chapter Thirteen

Preparing for the written work

The true purpose of the Essays

The Essays and Written Test are prime factors in the Reviews;
their purpose is to test your ability to communicate in writing
through scrutiny of the following aspects.

(a) The way in which the content relates to the question – this
includes the factual content and the way in which various
aspects of the subject are analysed, compared and contrasted
and conclusions reached.

(b) The structure – there must be a clearly identifiable *introduction*
displaying the main aspects of the topic; a *development*, in
which the topic is considered in detail and the main and
subsidiary arguments set out; a *conclusion*, in which the
threads of the discussion are drawn together and conclusions
reached.

(c) The quality of the English – this relates not only to grammar
and spelling, but also to sentence and paragraph construction
and style.

Perfecting a technique

The Reviewers look upon candidates as potential senior managers
who will be required to present reports or advice to laymen, so a
reasonably high standard is expected. Past inability to express
ourselves clearly (except to other engineers) has contributed to a

lack of public confidence. It is worthwhile therefore devoting some time to perfecting a technique; there are four complementary ways in which this can be achieved.

(*a*) Attendance on a course — there are many throughout the country, but look at the format carefully; you need to consider your requirements in detail:
- is a concentrated four-day course better than 26 weekly evenings?

- is your weakness the use of language, grammar and syntax or breadth of knowledge?

- would it help if your practice essays were marked?

Thomas Telford Training has a very good distance learning package which enables you to work around your normal work commitments — particularly useful if you are site based.

(*b*) A selective programme of reading — this includes publications relating not just to technical matters in which you are involved but to your profession, management generally and the political, financial and environmental framework within which you work.

(*c*) Development of a critical awareness in the workplace, not merely of your own involvement but of the whole environment within which you operate.

(*d*) Suitable practice under 'examination conditions'; this is by far the most important. As an invigilator, it is clear to me that too many candidates are *physically* unfit for the Essays, i.e. they cannot write for an hour and a half without discomfort. After all, when was the last time you wrote continuously for that length of time — during your finals? If you are thinking about cramp and discomfort, you cannot be concentrating on what you are writing!

The first Essay or the Written Test is set from your submission; it is expected that you will know at least enough to answer it. (If not, the Reviewers have found you out — your Reports have misled them!) Do read the question and make sure you understand what the Reviewers want; it helps if you underline the keywords. Do you understand what is required by 'Discuss', 'Describe',

'Comment' etc.? But remember, unlike much of what you have written over the last several years, this is not primarily a test of knowledge; it is a test of how well you can communicate your knowledge to a layman (that is, someone of equivalent intelligence but without your specific knowledge). In this respect it differs markedly from most, if not all, of the written communication you carried out during your academic education.

Do not start the actual writing without adequate planning. For 1000 words, 20 minutes should give enough time to construct a plan. This may seem excessive, but, with practice, most people can physically write 1000 words in less than an hour, leaving a few minutes to check the product at the end. At the start of an hour and a half, 20 minutes will seem like a lifetime – all your instincts will be screaming at you to start and everyone around you will be scribbling frantically. But control yourself, secure in the knowledge that previous experience under similar conditions has made you confident you can write sufficient in the remaining time.

Planning is itself assessed – any drafts, notes, plans etc. should be inside the cover of the Essay book if you are to receive any credit for your planning; this is why no scrap paper is allowed. Once you have decided on a plan, stick to it and finish it – your examiners will not be impressed if you cannot even fulfil your own plan. Monitor progress by deleting each idea once it has been incorporated.

How do I plan an essay?

Ideas will spring to mind in a very haphazard way as you consider the question. It is important to jot down every idea in note form as it occurs, however random and irrelevant it seems. The brainstorming after the initial blank panic at the start will, generally speaking, start with the answers to the question, followed by 'less important' material to provide enough to write 1000 words. The temptation is to follow the same logical process in writing your essay, answer the question in the first few pages and then proceed to justify the answer; the result is that the essay tails away. It is better to start by filling in the background first, arriving at the answer at the end – i.e. invert the list. In this way you lead the reader to a firm conclusion. But beware! Do not pad out the start with irrelevancies; how many times have I seen a Reviewer's comment somewhere around page three, 'Essay starts here' – not a good omen!

Once all your ideas are recorded, you can assemble them in a logical order; there are several ways of doing this:

(a) simply numbering each idea in a sequence and then rewriting in the correct order – the danger here is that it is relatively easy to miss one out,

(b) lettering (or using a colour code) to identify each idea with its appropriate paragraph and then numbering each within the paragraph, finally rewriting in the correct sequence,

(c) spider (spoke) diagrams where ideas radiate outwards from the central question, each being assigned to the most appropriate 'leg' – each leg subsequently becomes a paragraph,

(d) 'noughts and crosses framework', achieved by dividing the essay into three parts, each part containing three paragraphs. Each idea as it occurs is put into the most appropriate 'box'. This is a highly structured technique which requires practice, but is very useful for writing short reports.

(e) I have also come across another method known as 'mind maps', used by the Open University and on which a book has been written. It is a more sophisticated 'spider' system.

Use the system with which you feel most comfortable.

Do not get carried away by one particular aspect to the detriment of others; try to achieve a *balance* in both content and depth of treatment. Planning methods (d) and (c) and, to a lesser extent, (b) clearly highlight any such imbalance since one square or leg will become overloaded. If this happens, reconsider the distribution, consider adding another paragraph or perhaps you are entering into too much detail on this aspect.

Planning gives you the structure of the body of the Essay, especially the paragraphs. A paragraph is a collection of sentences all on one particular theme. Change the theme – start a new paragraph. The use of paragraphs demonstrates control over ideas and their expression. A good Essay contains paragraphs of variable length, each containing material on one aspect only. The paragraphs themselves should follow a logical sequence, avoiding abrupt changes in direction which disconcert the reader.

The book provided by the Institution for writing your Essay has a rather unexpected form; the back cover is A3 folded in. The covers themselves are different colours for different purposes and the invigilator will ask you to check that you have the correct one for your particular needs. But the purpose of this A3 cover is for you to carry out the brainstorming on the left-hand side and prepare the plan on the right-hand side. Left open, your plan is then visible all the time you are writing, to the right of the main booklet. Not so convenient for left-handed people, but still useful. (See the diagram at the end of Chapter Nine, which is similar.)

Deciding on a format

The Reviewer will also consider how effectively you introduce and conclude the Essay. The opening sentence must attract interest and make the reader want to read on; a short, sharp first sentence is the most effective. It can be a good idea to rewrite the question and outline your approach to it — this helps to ensure that you do actually answer the set question and not the one you thought (or hoped) it was.

There must be a conclusion, a 'summing-up'; even if you are short of time, at least a one-liner. Avoid dullness and assumptions; do not introduce further information; do make sure that your development does actually lead to the conclusion you have reached! The conclusion must be positive — do not allow the essay to peter out.

The Reviewers are looking for understanding supported by facts. Do not be afraid to express original ideas and opinions (as long as they are sensible and you justify them) rather than regurgitating stereotyped popular views. Present points in an orderly, uncomplicated way to demonstrate logical, clear and, if possible, original thought. Substantiate each point with facts or figures, wherever possible taken from your own experience. You have met your Reviewers by this stage — do not fall into the trap of subconsciously writing for them, knowing as you do that they are already familiar with the subject – the result will be superficial and unacceptable.

Clarity and presentation

Make the script look organised and authoritative. Practise handwriting if yours is rather scruffy and difficult to read. In any case, when was the last time you wrote for an hour without a break – you need the practice. One useful method is the Quarterly Report; of the same length but different format, once you have decided on a plan you should be able to write each one in an hour (though I have yet to find anyone who has!). Try to keep the handwriting consistent throughout so that, once accustomed to your style, the Reviewer can read it easily. Indent paragraphs in a consistent way – begin on the next line (no space between paragraphs) about 3 cm from the left-hand margin.

Remember that the Reviewer is reading your work in their own time and may well be tired; make it clear, readily understandable and enjoyable, but avoid making jokes – they will inevitably be taken the wrong way!

Avoid lapsing into jargon or slang; do not use words like rebar, shutters, dumper or lab. This is not easy, because we use both jargon and slang every day in discussions with other engineers, but these are not considered appropriate in the written word.

Under no circumstances use several words where one would do because you think the script may be too short; this will only draw the Reviewers' attention to the deficiency and probably cause annoyance. If you use abbreviations or acronyms, write them out in full, followed by the acronym in brackets, on the first occasion on which they appear in your Essay.

Practice makes perfect

To be successful in the Essay or Written Test it is vital to practise under the supervision of someone (not necessarily another engineer) who can comment not so much on the content but on the 'readability' of your work. They need to be asked such questions as: 'Do you see what I am getting at?' 'Can you follow my argument?' 'Does it make sense?' 'Did you enjoy reading it?' Such a person need not be an engineer; in fact sometimes it is a positive benefit that they are not familiar with the subject that you are writing about. They will ask questions which cause you to consider whether or not you have actually written what you intended, or whether there is a better way of explaining that particular matter.

Having watched a large number of candidates writing and discussed the problems with them and the Reviewers, it is absolutely certain that there is no substitute for practice – not just in writing essays, but in writing them under severe time constraints. Anyone can write an acceptable essay in a fortnight; it requires skill, time management and clear, quick thinking to achieve an adequate result in an hour and a half. I become concerned when I visit essay groups at the amount of time and effort spent collecting knowledge – generally far too much to regurgitate in the permitted time. All that many of these groups are in fact achieving is to burden their members with an additional problem – too great a choice of available information! By all means read and discuss around the subjects, but then distil this information into key points, which can be introduced into many of the questions. It is my considered view that most essay groups would be much more effective if they concentrated more on practice than on collecting information.

The other concern I have is the predilection for 'model' answers. In one extreme example, a candidate with a background entirely in water engineering answered a question using examples from highway work. Obviously he had swatted up model answers and was no doubt disconcerted when he failed on Knowledge and Relevance! Even within a group from the same organisation, the answers will be different, reflecting the differing experience of the members. Model answers may give you a lead and a general shape to an answer, but you must put your own perspective on them.

You need a working knowledge of the whole range of civil engineering, not merely a high degree of specialist technical knowledge, so that you can relate your expertise to the environment in which we work. You also need to immerse yourself in well written, well structured papers, magazines and books so that you subconsciously absorb good style and techniques.

What if my first language is not English?

The language of the Institution is English, a fact recognised and accepted worldwide. You will therefore be expected to have a reasonable facility with it. You are trying to prove that you are a professional engineer, and as such, *any* reports or documents which go out from your organisation would be expected to be of a consistently high standard. So your Reports and submission must be perfect. As a responsible engineer, you are expected to have taken any necessary steps to ensure this, whether English is your first language or not.

The standard expected for the written work in the afternoon of the Review is different. The documents state 'a reasonable first draft'. In other words, not perfect, but tidy and logical, capable of being edited by someone without your technical background. A few crossings out might be acceptable, but great circles around paragraphs to move them elsewhere are not – they demonstrate a lack of foresight and a muddled mind. Again, your curiosity should have been aroused by the wording in the official booklets, where the requirement states 'in acceptable English'. 'Acceptable' is perhaps an unexpected adjective, but it allows the Reviewers the freedom to make a judgement – acceptable in the context of the background of the candidate.

This proviso applies also to dyslexia, provided that you are able to inform the Institution at the time of submission by supplying a certificate defining your problem. There is no dispensation on the time allowed, but all such written work by dyslexic candidates is automatically referred to a special panel with access to appropriate medical advice.

Chapter Fourteen
The aftermath

 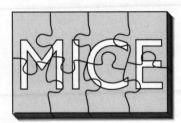

Why do the results take so long?

You have completed the jigsaw, to the very best of your ability, by about the second week in October or May at the latest. You now face a prolonged wait for the result, which will be published either just after Christmas and before New Year or on 30 June. It is perhaps difficult for candidates to understand why the assessment process takes so long.

When you leave the essay room, the system really swings into action. The literary efforts, amounting to as many as forty scripts each day for a month, are sorted, parcelled up and posted (generally that same night) to the least experienced Reviewer (who probably came out to make the initial introduction at the start of the interview). They will read and mark your work before posting it off to the Lead Reviewer, who in turn will read the essay(s) and then confirm the marking. At this stage they will probably confer by phone either to compromise where there are discrepancies in the marking, or to discuss the final completion of their paperwork about you, most of which they try to complete in the short time between each interview. Obviously they now have to add the result of the written work and confirm the tentative conclusion which they arrived at after your interview. Remember that they are all doing this in their own time and fitting it into an already busy schedule of work so there are bound to be some delays. Also remember that they are under some pressure, since the rules state categorically, 'If there is any doubt whatsoever, the candidate must fail'.

Your written work, together with the Reviewers' forms, their verdict and a draft of the letter to be sent to you if you have not been successful, are then all parcelled up and sent to the Reviews Office. As between 600 and 800 results arrive, the office staff prepare all the letters for signing by a senior officer, and take the samples which will be put before expert panels for review as part of the quality assurance processes. At this stage, there may also be dyslexic candidates whose written work is automatically reviewed by a panel. These panels are specially convened and again are dependent on a number of volunteers fitting mutually convenient dates into their diaries.

Meanwhile, the staff file all the paperwork and examine any responses to the announcement published in *New Civil Engineer* some time before, where the membership is asked whether anyone has any reason why the listed names (including yours) should not be admitted to the Institution, provided they are successful at the Review. The draft failure letters are also being examined for any inconsistencies before being printed and signed.

I hope you can now see that everything possible is done to keep the suspense time to a minimum. This is why you will definitely not get any reply to a query about the results. All that such a phone call or enquiry will accomplish is to delay the process for everybody – so do not be tempted.

The unsuccessful review

There are all sorts of myths about the result envelope; people are convinced that you can tell the result from the size of the envelope. Do not be misled, just open it. If you see the dreaded words 'I regret to inform you . . .' that is probably all you will read for some time – all that effort and hard work was in vain. This is a devastating blow. There is little anyone can do to reduce its impact; the Reviewers are urged to tell you what you did correctly, what experience was good and go on to tell you why they reached their decision, but the initial shock will prevent you from looking at this rationally.

You will then go on to read the reasons and will become increasingly disillusioned. It is highly likely that you will disagree with what you think the letter says and feel that it is unfair. The temptation will be to march down the corridor and berate your Supervising Civil Engineer or line manager about the inadequacies of the Institution or pick up the phone and talk belligerently to your Regional Liaison Officer. Worse, you may even be tempted to write a strong letter to the President! The one thing you definitely should never do is phone the Reviewers; this is considered totally unprofessional and completely out of order.

Recovering the situation

I strongly counsel caution. Wait until your initial surge of disappointment, even anger, has dissipated. You will then feel utterly dejected. It is at this stage that you need wise counsel, something which the Regional Liaison Officers are in a very good position to provide. They see many letters (you have seen only this one and your colleagues hopefully very few more) and they know most of the Reviewers and their styles of writing. They also have access to advice and guidance from their colleagues on matters of detail with which they might not be familiar. So they will be able to give you an insight into what has actually been said – 'reading between the lines' as it were.

The phrase which appears most frequently is 'You failed to demonstrate . . .'. Your reaction may well be 'But it's obvious!'. Is it? Or are you expecting the Reviewers to infer something which the regulations state has to be *demonstrated* and which has not actually been spelt out in your Reports? You *know* what you were trying to write; they can only see what you actually wrote down. Did you actually say what you meant?

You may feel that you were not given adequate opportunities

during the interview to display your expertise; you must ask the question 'Why?'. Had the Reviewers come to the conclusion that it was not worth pursuing this matter? Your interview should not have been a cross-examination – poking relentlessly at a weakness may only cause a candidate to become defensive and is unlikely to allow them to present themselves favourably, so most Reviewers change the subject. Could you, in fact, have demonstrated the point in question or have they actually identified a real weakness?

Very often, the letter concludes by suggesting that before you submit again, you should seek advice from senior engineers, including your Regional Liaison Officer. This is good news, because it suggests that your Reviewers actually believe that you stand a fair chance of success if you just rectify the things which they have identified.

Rectifying any problems

There will be one of two outcomes from this analysis of the detail of the letter – either you did 'fail to demonstrate' or the Reviewers actually did identify a weakness. The former can fairly readily be rectified in the next submission, while the latter will take some time. In extreme cases, it may be that they have exposed fundamental weaknesses, either personal or in your experience, which make another attempt unwise. They may suggest that it may be more suitable for you to divert to a different, more appropriate Review, certainly for the time being. This is a formidable personal decision to have to make and I urge anyone in this difficult situation to discuss it with as many informed people as they feel able. Do not keep mulling it over yourself – you will not resolve the problem, it will merely loom larger.

Any small shortfall must be put right by further experience; it may be that the Reviewers have given an indication of how this could be achieved. You need to talk it through with key personnel in your organisation, perhaps engineers who sponsored you, to identify opportunities for suitable experience which might become available soon to rectify this weakness and how you could take advantage of them. Perhaps the problem was with the written work, in which case it may be advisable to enrol on an appropriate course. This could be one of the many specifically aimed at the Reviews, or it could be a more fundamental course on the use of the English language.

The revised submission

You need to review your entire submission to see whether the various documents can be adjusted to cover the weaknesses perceived last time. You will have had, in the interim, further experience, which might be more appropriate for the Project or Expertise Report; in any case, your Experience Report will have to be adjusted so that you can include this extra experience.

Every Review is a complete entity. Do not make the fundamental mistake of believing that you have passed most of the Review and now only need to rectify the few shortcomings, either perceived or real. You are taking the whole Review again – with one important difference. This time the Reviewers will have an additional document – a copy of your failure letter from the Institution. This puts you at a slight disadvantage, so you must get back to the status quo as quickly as possible.

To do this, I believe you must answer the letter which has been copied to the Reviewers. I think it entirely appropriate that you write to each of them (as part of your revised submission), explaining precisely what you have done, in the way of additional experience, Continuing Education and Training to rectify weaknesses or in rewriting the submission and amending the supporting documents, to rectify your 'failure to demonstrate' last time. Otherwise they will not be aware of your efforts and you will have to clarify your actions during your interview, which just shortens the time available for them to find out whether you now comply with the requirements and really are a professional engineer.

Chapter Fifteen

Mature Candidate Review

Eligibility

This route is almost worth a book on its own. In an ideal world, perhaps a separate volume would be the best solution, because it is so different from all the other ICE Reviews. Keeping it separate would accentuate the differences. This is why I have left this chapter until the end. Like the other chapters, however, this one again concentrates on the philosophy and purposes; you are expected to pick up the rules and regulations from the Institution's publications, in this case ICE 104, *Routes to Membership – Mature Candidate Review*.

The Mature Candidate Review provides a route to an appropriate grade of membership only for those *without the required academic qualifications*. There is a mistaken belief that it is the route for all those whose age exceeds the minimum requirement; this is not the case and in fact would subject many candidates to a more onerous review than is necessary through the more conventional Reviews.

There are three mature routes – to Chartered Engineer, Incorporated Engineer and Engineering Technician. All three follow the same stages and have similar formats, but the detailed requirements vary. So it is important to choose the best route for you and then read the instructions carefully.

The very first thing to do is to decide which is the most appropriate grade of membership. Compare yourself with the criteria referred to in Chapter Two and make an honest judgement about yourself. One of the reasons why so many stages have been introduced into this route is that far too many people seem to believe that merely because they have a lot of experience in civil engineering and have reached maturity, they have a chance of success through the Mature Route. This is just not true!

Far better to persuade yourself at the outset of the most suitable and realistic goal than to aim too high and be told, after a large amount of effort and a considerable period of time and several interviews, that your experience has not developed the required levels of ability. There is no room for potential, as there might be for a young engineer; at the minimum age of thirty-five, you must have, to all intents and purposes, become a professional engineer.

The second thing to do before you start in earnest is to make quite certain that your academic qualifications are not acceptable and that you therefore cannot utilise the conventional routes. To some extent, this depends not only on what your qualifications are, but also on exactly when, and perhaps where, you achieved them. The rules are complex and you need to get informed advice and guidance from the Education Department at the Institution. Many overseas qualifications are (or can be) ratified, so even if your degree or diploma is not apparently eligible, do check. You could save yourself a significant amount of effort.

Why is this route so different?

Most engineers follow a progression towards their professional qualification, through academic education, followed by structured training and responsible experience. If you have been honest in your assessment of yourself against the requirements of Chapter Two, you should be entirely satisfied that you have reached the required level of responsibility, but without following the normal progression.

So for this Review, you have two things to prove – first, that you are operating at a level of responsibility commensurate with that of the professional qualification you seek and, secondly, that you have achieved a standard of technical and academic competence comparable with that possessed by an academically qualified engineer in the same position of responsibility. This is the fundamental difference between this review and all the others. In the UK, the Engineering Council is responsible for setting the minimum standards for academic achievement. They do, therefore, take an active part in your review.

It is this academic requirement which seems to cause the greatest difficulty in the progression, mainly, I think, because candidates concentrate exclusively on proving their professional responsibility. It is, after all, probably their workplace competence which motivates candidates to become professionally qualified in the first place.

By and large, your c.v. and Experience Report are the vehicles to prove that you are operating at a sufficiently high level of responsibility, while your Technical Report will demonstrate your understanding of technical principles. It is vital that you keep this distinction in mind throughout your preparation of the documents.

Choosing your Lead Sponsor

You are asking this person to fulfil an onerous task, which will involve them in making the decision as to whether you stand a reasonable chance of success and then reading and advising during the preparation of your documents and attending meetings with you. To fulfil these duties, they must not only have an up-to-date knowledge of the requirements of the appropriate grade of membership, but must also understand and appreciate the essential differences between the Reports you are about to prepare and those prepared by conventional candidates. This latter aspect is vital; I have counselled several unsuccessful candidates who have received advice, given in the utmost good faith by established Reviewers of conventional candidates, but which has proven incorrect for this route.

I therefore strongly suggest that at an early stage, after you have selected a Lead Sponsor, you both arrange a meeting with the

Institution's Regional Liaison Officer for your area, so that you can be sure that your proposed plan of action fully conforms with the system and that all parties are absolutely clear about what is to be demonstrated.

Preparing the Experience Report

Your Report on practical experience has to cover at least fifteen years' experience and yet is no longer than the similar Report for the Chartered Professional Review, where candidates usually have about half that length of experience. So you must use every device available to present sufficient information to the Appraisers to enable them to make a preliminary judgement on whether you are a professional engineer. Fortunately, you do also have other documents to assist – your c.v. and a copy of the Core Objectives or Schedule of Sectors and Activities.

The keyword for the c.v. is 'current'; the Institution is asking you to use the greater part of this to outline your *current* role and responsibilities, with just a brief sketch of the career which enabled you to reach that level of competence. It therefore seems sensible to write it in reverse chronological order, first describing what you are doing and then briefly explaining how you got there.

The Core Objectives in ICE 180 have been formatted in such a way that you will have approximately half a page to describe how each objective has been achieved. By carefully completing this in conjunction with, and at the same time as, the report on Practical Experience, it is possible to cover your total experience adequately and to draw out the key factors in your experience which have compensated for your lack of formal education and training. But this process takes time! It is not something which can be done at the last minute.

I suggest that you take copies of the Objectives or relevant Activities (of which there are likely to be a considerable number after so much experience) and make notes on each one of how and to what level it has been satisfied. These notes will enable you to compile a detailed inventory of your earlier experience, to which you can then add the work which demonstrates all the qualities of a professional engineer. It is then a question of severe editing and fitting all the information into the two documents: the Activities or Objectives and the Experience Report.

Throughout the preparation of these three components, you must keep in mind the requirements for a conventional candidate at both the Training Review and the Professional Review. In general, your c.v. will tell the Institution what experience you have been exposed to, the report on Practical Experience and the Core Objectives or Activities will tell them all the benefits you gained from that experience and how you developed the competencies of a professional engineer. It is vital that you write these documents with these targets in mind, otherwise you will follow a long list of people who have informed the Institution only of what they have done, but failed to tell them of the benefits they gained.

Much later on in the process, when your application is referred to the Engineering Council, you will be asked to write for them a brief explanation of why you believe you should be allowed to be a professional engineer. Surely it makes sense to take their description of an engineer (at the appropriate grade) and use this to describe yourself through your work? In this way you should cover all the aspects which the Engineering Council will be seeking. It is worth considering this, and perhaps producing a draft, at this early stage because, again, it will focus you on exactly what you are trying to prove.

Producing the synopsis and Technical Report

Do not produce the synopsis before you at least have a draft Technical Report. Like the documents in the paragraph above, these two cannot be written in isolation, but are totally dependent upon each other and must be written together. I hate to think of the number of candidates who have written and submitted a synopsis and then found they could not write the Technical Report to back it up! This synopsis is critical and is frequently a stumbling block for a large proportion of candidates. So again, it needs careful construction, targeting the criteria laid down in ICE 104, and remembering that it is not a resume, summary or overview but a *synopsis*. Published research papers are a good place to find out how to write a synopsis if you are not sure.

The criteria for your Technical Report are laid down in some detail in ICE 104. The primary thing to realise is that, unlike the Project or Expertise Reports for conventional candidates, you are not trying to prove your abilities as a professional engineer, but that you have the same understanding of the technical principles behind your particular area of competence as they have.

Two of your three Reviewers will probably be academics, one of whom is a member of the Institution of Civil Engineers, the other from any of the Institutions which are affiliated to the Engineering Council. This gives you a good indication of the understanding of technical principles which is required – really fundamental, understood by all disciplines of engineering. It is no good merely demonstrating that you are familiar with Standards and Codes of Practice or comfortable with various design methods, still less is it enough to show your expertise in construction management. You *must* demonstrate a clear understanding of the fundamental behaviour of the materials and structural forms on which you have chosen to be examined. It is not enough that you know how to do the calculations, you must demonstrate that you fully understand the function that the calculations are performing.

This probably means that you are going to have to do some work before you write your Report. For example, candidates have frequently found it advantageous to attend the final year of an appropriate course at college or university before submitting. Nearly all successful candidates have found it necessary to read in depth and study round their subject area in order to be able to demonstrate an adequate understanding. In the same way that students during their final academic examinations are required to memorise and quote various formulae and equations which, under normal work situations, they would look up in an appropriate textbook, you will be expected to be able to do something similar during your Review.

Summary

The burden of proof for the Mature Candidate Route is described by the Institution as stringent. British professional engineering qualifications are frequently perceived within the European context as lacking in formal academic achievement, so there is constant pressure on the Engineering Council to prove otherwise. Those without any adequate academic qualifications can therefore expect to be examined in some considerable depth. If the process is tackled logically and steadily, with advice and guidance at every stage, and you have the will to spend time on achieving a thorough understanding of the technical principles, then there is a very good chance of success for those who *are* professional civil engineers in all but qualification.